Praise for Violet Moller's

THE MAP OF KNOWLEDGE

"Fascinating. . . . A picturesque tour of a series of fabulously wealthy civilizations. . . . Moller brings the wonders of the medieval Muslim empires vividly to life and you're left yearning for more."
—*The Times* (London)

"Unusual and well-crafted. . . . An impressive, wide-ranging examination of what might be called premodern intellectual and cultural geography."
—*Publishers Weekly*

"Moller enlivens her history with stories about young scholars who dedicated their lives to preserving these valuable texts. . . . A dramatic story of how civilization was passed on and preserved."
—*Kirkus Reviews*

"An epic treasure hunt into the highways and byways of stored knowledge across faiths and continents."
—John Agard, poet and judge
of the Royal Society of Literature 2016 Jerwood Award

"An exceptionally bold and important book."
—Daisy Hay, author of *Young Romantics*
and judge of the Royal Society of Literature 2016 Jerwood Award

Violet Moller

THE MAP OF KNOWLEDGE

Violet Moller is a historian and writer based near Oxford, England, with a PhD in intellectual history from the University of Edinburgh. She won the Jerwood Award from the Royal Society of Literature in 2016 for a first major work in progress for *The Map of Knowledge.*

www.violetmoller.com

THE
MAP *of* KNOWLEDGE

A THOUSAND-YEAR HISTORY OF
HOW CLASSICAL IDEAS WERE LOST AND FOUND

Violet Moller

Anchor Books

A DIVISION OF PENGUIN RANDOM HOUSE LLC

NEW YORK

FIRST ANCHOR BOOKS EDITION, APRIL 2020

Chapter opener illustrations by Anna Woodbine.

The Library of Congress has cataloged the Doubleday edition as follows:
Name: Moller, Violet, author.
Title: The map of knowledge : a thousand-year history of how classical ideas
were lost and found / Violet Moller.
Description: First American edition. | New York : Doubleday, 2019.
Identifiers: LCCN 2018054987
Subjects: LCSH: Learning and scholarship—Mediterranean Region—History—
Medieval, 500–1500. | Mediterranean Region—Intellectual Life—History. |
East and West.
Classification: LCC AZ231 .M65 2019 | DDC 001.209182/2—dc23
LC record available at https://lccn.loc.gov/2018054987

Anchor Books Trade Paperback ISBN: 978-1-101-97406-3
eBook ISBN: 978-0-385-54177-0

www.anchorbooks.com

Printed in the United States of America
10   9   8   7   6   5   4

To L, E and S,
my three little stars

CONTENTS

Preface

IN EARLY 1509, the young artist Raffaello Sanzio (1483–1520) began painting a series of frescoes on the walls of Pope Julius II's private library, deep inside the Vatican. Next door, in the Sistine Chapel, Raphael's great rival, Michelangelo, balanced on a huge scaffold hundreds of feet in the air, was painting a monumental image of God giving life to Adam onto the ceiling. The Renaissance was in full swing in Rome and, under the patronage of Pope Julius, the great city was being returned to the glory of its ancient imperial past. Raphael's frescoes on the four walls of the Stanza della Segnatura illustrated the four categories of books that were shelved below them: theology, philosophy, law and poetry. In the philosophy fresco, which we now call *The School of Athens*,* Raphael painted three huge vaulted arches receding into the distance, with statues of the Roman gods Minerva and Apollo on either side and broad marble steps leading down to a geometrically tiled floor.[1] The architecture is decidedly Roman—bold, imperious, monumental—but the culture and ideas represented by the fifty-eight figures carefully grouped across the painting are emphatically and almost without exception

---

* Not a school in the modern sense, but a loose circle of individuals with similar academic interests and, in this case, a tradition of study lasting several hundreds of years.

Greek; it is a celebration of the rediscovery of ancient ideas that were central to the intellectual milieu of sixteenth-century Rome. Plato and Aristotle stand in the very centre, under a huge arch, silhouetted against the blue sky, which Plato points up to, while Aristotle gestures to the earth below him, neatly representing their philosophical tendencies—the former's preoccupation with the ideal and the heavenly, the latter's determination to understand the physical world around him. The full scope of ancient philosophy as inherited by Italian humanism is triumphantly rendered in glowing colour.

No one knows exactly who all the other figures in the fresco are, and arguments over their identities have kept scholars occupied for centuries. Most people agree that the bald man in the front right, busy demonstrating geometrical theory with a compass, is Euclid,[2] while the crowned man next to him, holding a globe, is certainly Ptolemy, who at this point was far more famous for his work on geography than astronomy.* All the figures identified lived in the ancient world, at least a thousand years before Raphael began painting the fresco—except for one. On the left of the painting, a man wearing a turban is leaning over Pythagoras' shoulder to see what he is writing. He is the Muslim philosopher Averroes (1126–1198)—the single identifiable representative of the thousand years between the last of the ancient Greek philosophers and Raphael's own time, and the single representative of the vital, vibrant tradition of Arab scholarship that had flourished in this period. These scholars, who were of various faiths and origins, but were united by the fact that they wrote in Arabic, had kept the flame of Greek science burning, combining it with other traditions and transforming it with their

---

* In the Renaissance, scholars mistakenly believed that Claudius Ptolemy the astronomer and geographer was a member of the Ptolemaic dynasty, who ruled Egypt from 305 to 30 BC.

own hard work and brilliance—ensuring its survival and transmission down through the centuries to the Renaissance.

I studied Classics and history throughout my time at school and university, but at no point was I taught about the influence of the medieval Arab world, or indeed any other external civilization, on European culture. The narrative for the history of science seemed to say, "There were the Greeks, and then the Romans, and then there was the Renaissance," glibly skipping over the millennium in between. I knew from my medieval-history courses that there wasn't much scientific knowledge in Western Europe in this period, and I began to wonder what had happened to the books on mathematics, astronomy and medicine from the ancient world. How did they survive? Who recopied and translated them? Where were the safe havens that ensured their preservation?

When I was twenty-one, a friend and I drove from England to Sicily in her old Volvo. We were researching Graeco-Roman temples for our third-year dissertations. It was a great adventure. We got lost in Naples, hot in Rome, we were pulled over by the police and asked out on a date, we gaped at Pompeii and ate milky balls of buffalo mozzarella in Paestum, and finally, after weeks on the road and a short ferry trip across the Straits of Messina, we arrived in Sicily. The island immediately felt different from the rest of Italy: exotic, complicated, compelling. Its layers of history enveloped us; the marks left by succeeding civilizations, like strata in a rock face, were striking. In Syracuse Cathedral, we saw the columns of the original Greek Temple of Athena, built in the fifth century BC, still standing 2,500 years after they were erected. We learned how the cathedral had been converted into a mosque in 878, when the city came under Muslim control, and how it became a Christian church again two centuries later, when the Normans took power. It was clear that Sicily had been a meeting point for cultures over hundreds of years, a place where ideas, traditions and words had been exchanged and transformed, where worlds had collided. The focus of our trip was the relationship between Greek and Roman religion and architecture, but the contribution of later cultures—Byzantine, Islamic, Norman—was remarkable. I began to wonder about other places that had played a similar role in the history of ideas, and how those places had developed.

These questions resurfaced when I was researching my PhD on intellectual knowledge in early modern England, viewed through the library of Dr. John Dee (the man Elizabeth I called her philosopher). A strange and captivating character, Dee was my constant companion for several years. He took me on an unforgettable journey through the intellectual world of the late sixteenth century. During his extraordinary career, he amassed the first truly universal collection of books in England, helped plan voyages of discovery to the New World, initiated the concept of a British Empire, reformed the calendar, searched for the philosopher's stone, attempted to conjure angels and travelled all over Europe with his wife, servants, several children and hundreds of books in tow. He also wrote extensively on a wide range of subjects: history, mathematics, astrology, navigation, alchemy and magic. One of his most significant achievements was helping to produce the first English translation of Euclid's *Elements*, in 1570. But where had this text been and who had looked after it in the 2,000 years between Euclid writing it in Alexandria and Dee publishing it in London? Studying the catalogue Dee made of his library in 1583, I noticed that a great many of his books, especially those that touched on scientific subjects, were written by Arab scholars. This tied in with the things I had seen in Sicily and gave me a taste of what had been going on in the Islamic world in the Middle Ages, expanding my view of history beyond the traditional Western scheme. I began to realize that the history of ideas is not constrained by boundaries of culture, religion or politics, and that, in order to fully appreciate it, a more far-reaching approach is necessary.

These ideas remained at the back of my mind, gradually crystallizing into a plan for a book that would follow ancient scientific ideas on their journey through the Middle Ages. As it is an enormous subject, I decided to concentrate on a few specific texts and plot their progress as they passed through the major centres of learning. With my focus on the history of science and, more precisely, "the exact sciences," three subjects were clearly delineated: mathematics, astronomy and medicine.* Within them, three geniuses stand out: in mathematics, Euclid; in astronomy, Ptolemy and, in medicine, Galen. Euclid and Ptolemy both wrote comprehensive surveys of their subjects—*The Elements* and *The Almagest*—but Galen was a more complex proposition. He wrote hundreds of texts, so I decided to concentrate on those that formed the medical curriculum in Alexandria, in addition to the general areas of anatomy and pharmacology. All three of these remarkable men defined the structure and content of their individual subjects. They created the framework within which future scholars would work for hundreds of years. Many of the theories of Ptolemy and Galen have since been disproved and replaced, but their influence and legacy is incontrovertible. Galen's theory of the humours still survives in traditional Tibetan medicine and also in modern complementary medicine. Ptolemy's survey of the fixed stars endured, as did his "idea that the physical world is dependable and can be understood with mathematics."[3]

In contrast, Euclid's *Elements* has stood the test of time, almost in its entirety. It was still being taught in classrooms in the twentieth century, and the geometrical theories it contains remain as

---

* In the ancient and medieval periods, scientific subjects fell under the umbrella of "natural philosophy"—any investigation into the physical world.

true and relevant today as they were in the fourth century BC. The same applies to Euclid's demonstrative method, which uses a concise technical vocabulary, suppositions and proofs (diagrams), and which has remained a template for scientific writing ever since. Euclid, Galen and Ptolemy pioneered the practice of science based on observation, experimentation, accuracy, intellectual rigour and clear communication—the cornerstones of what is now known as "scientific method."

When I began to research in earnest, I was surprised how neatly the story unfolded in front of me. The year 500 was an obvious moment to begin—a time when the intellectual traditions of antiquity were evolving into those of the Middle Ages, when scholarship was entering a different era. Subsequent chapters each centre on a different city, first of all doubling back to Alexandria to see when and how the texts were written. From here, they were dispersed across the Eastern Mediterranean to Syria and Constantinople, where they remained until the ninth century, when scholars from the new city of Baghdad, capital of the vast Muslim Empire, began seeking them out to translate them into Arabic and use the ideas contained in them as the foundation for their own scientific discoveries. Baghdad was the first true centre of learning since antiquity, and over time it inspired cities across the Arab world to build libraries and fund science. The most important of these was Córdoba, in southern Spain, ruled over by the Umayyad dynasty, under whose patronage the works of Euclid, Ptolemy and Galen were studied and where their ideas were questioned and improved upon by generations of scholars. From Córdoba, they were taken to other cities in Spain and, when the Christians began to reconquer the peninsula, Toledo became an important centre of translation and the place where they entered the Latin, Christian world.

This was the major route the texts took, but there were other places in the Middle Ages where ancient Greek, Arabic and Western culture collided. Salerno, in Southern Italy, was a place where medical texts (in Arabic, but derived from Galen) were taken from North Africa and translated into Latin, and, as a result, Salerno became the centre of European medical studies for centuries, playing a vital role in the dissemination of medicine. Then, in Palermo, Ptolemy and Euclid take centre stage from Galen, as scholars translated copies of *The Elements* and *The Almagest* directly from Greek to Latin, bypassing the Arabic versions in the hope of achieving greater accuracy. The three divergent strands came together in Venice, where manuscripts began to arrive in the last half of the fifteenth century, ready to be printed for the first time.

There were, of course, other cities I could have included, but sticking to those in which copies of the key texts were studied and translated seemed the best way to avoid getting lost in this huge story. Choosing them threw up some interesting questions about what constitutes a centre of learning. Constantinople was a major repository of ancient texts, but not somewhere that science was studied with any degree of originality or rigour. Nor was it a place where translation (and therefore transmission) happened on any kind of scale and, for these reasons, it only features in a supporting role—the place to which scholars and caliphs came when searching for copies of Euclid, Ptolemy and Galen. The city on the Golden Horn might have taken over from Alexandria in terms of power and status, but it was a pale shadow when it came to scientific learning—a centre of preservation rather than innovation. Toledo, Salerno and Palermo were the places where Arabic culture came into closest contact with Christian Europe, but there was also a degree of exchange in Syria during the Crusades. I have not discussed this in much detail, however, because there is no evidence that *The Elements*, *The Almagest* or Galen's major works were among the books translated there.

While the underlying narrative for this story was easy to follow, finding a way through the dense, tangled undergrowth of manuscript history was not. Because they were so significant, several editions of each text were produced—teasing out their relationship to one another and finding a clear path through was often challenging. Until the introduction of the printing press, every single text was copied out by hand, so each one was different, with its own peculiarities and mistakes. The study of complex textual traditions is a discipline all of its own within history, and not one to which I can claim expertise. In order to stay true to the narrative, I have had to be selective and produce simplified versions of the rich manuscript histories of these great books.

For me, the history of ideas has always been the most fascinating aspect of our past. Discovering how people approached the fundamental questions about our planet and the universe, how they passed their theories on to future generations and expanded the frontiers of intellectual knowledge, is compelling. Much of this type of history is obscured in erudite books on the shelves of research libraries, but this should not be the case. By taking a broad view and writing about the characters and stories, rather than focusing on the scientific content and historical minutiae found in academic books, it is possible to bring the history of ideas to life. For example, understanding Ptolemy's model of the universe is above and beyond anyone without detailed knowledge of astronomy, but appreciating its importance and following its progress is both meaningful and fascinating. Doing so takes us on a sweeping journey through the Middle Ages, zooming in on certain places at certain moments to discover exactly how and why scientific ideas were transmitted and transformed. In this way, the boundaries of the traditional Western historical narrative are expanded by shining a light on the profound contribution of both the Islamic world and medieval Christian scholars, filling in the millennium between "the Romans" and "the Renaissance." This made it possible to include theories from other cultures which were gradually incorporated into the canon of mathematical, astronomical and medical thought. Ideas like the Hindu-Arabic numerals and positional notation system that came from India, via the Muslim Empire, and are used all over the world today.

When you step back and look at history from a wider angle, the intricate web of connections between different cultures comes into focus, giving us a broader, more nuanced and ultimately more vivid view of our intellectual heritage.

CELTIC LANDS

North Sea

JUTES

ANGLES

SAXONS

• York

Atlantic Ocean

Londinium •

JUTES

SAXONS

ARMORICA

FRANKISH KINGDOM

Tours
•

ALEMANS

BURGUNDIAN
KINGDOM

ALPS

CANTABRIANS

BASQUES

OSTRAGOTI
KINGDOM

SUEVIC
KINGDOM

CORSICA

• Rome

VISIGOTHIC KINGDOM

BALEARICS

SARDINIA

V
A
N
D
A
L

S

SICIL

S

Hippo
•

K
I
N
G
D
O
M

B E R B E R S

| 0 | 20 | 40 | 60 | 80 | 100 miles |
| 0 | 25 | 50 | 75 | 100 | 125 | 150 kilometres |

The World of Learning in 500AD

GEPIDAE

Black Sea

EASTERN

●Constantinople

Propontis

ROMAN

●Pergamon

EMPIRE

Athens●

●Ephesus

Antioch●

●Palmyra

PERSIAN (SASANIAN) EMPIRE

●Rhodes

●Damascus

diterranean Sea

Jerusalem●

Alexandria●

ARABIAN
DESERT

EGYPT

Red Sea

Nile

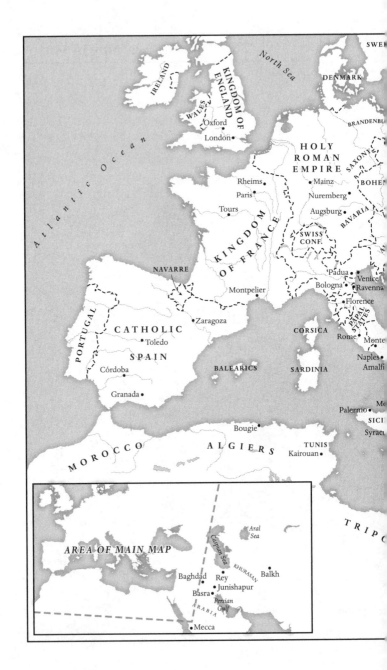

SWE[DEN]

North Sea

IRELAND

KINGDOM OF ENGLAND

WALES

Oxford
London

DENMARK

BRANDENBU[RG]

HOLY
ROMAN
EMPIRE

SAXONY

BOHE[MIA]

Atlantic Ocean

Rheims

Paris

Tours

Mainz

Nuremberg

Augsburg

BAVARIA

KINGDOM OF FRANCE

SWISS
CONF.

Padua • Venice
Bologna • Ravenna

NAVARRE

Montpelier

Florence

Zaragoza

CORSICA

PAPAL STATES

Rome • Monte

PORTUGAL

CATHOLIC
Toledo

SPAIN

Córdoba

Granada

BALEARICS

SARDINIA

Naples •
Amalfi

Palermo •
SICI[LY]
Syrac[use]

Me

Bougie

MOROCCO

ALGIERS

TUNIS
Kairouan

TRIPO[LI]

AREA OF MAIN MAP

Aral
Sea

Caspian Sea

KHURASAN

Balkh

Baghdad

Rey

Junishapur

Basra

Persian
Gulf

ARABIA

Mecca

The World of Learning in 1500AD

EUTONIC ORDER

NIC
R

LITHUANIA

AND

MOLDOVIA

GARY

WALLACHIA

KHANATE
OF CRIMEA

Black Sea

OTTOMAN EMPIRE

Constantinople

Propontis

Zara

Nisbis

Athens

Rhodes

Antioch

Damascus

diterranean Sea

Jerusalem

Alexandria

Cairo

SULTANATE OF EGYPT

ARABIAN
DESERT

Red Sea

Nile

| 20 | 40 | 60 | 80 | 100 miles |
| 25 | 50 | 75 | 100 | 125 | 150 kilometres |

ONE

The Great Vanishing

Greek scholars were driven out of the Greek world and helped to develop Arabic science. Later the Arabic writing was translated into Latin, into Hebrew, and into our own vernaculars. The treasure of Greek science, most of it at least, came to us through that immense detour. We should be grateful not only to the inventors, but also to all the men thanks to whose courage and obstinacy the ancient treasure finally reached us and helped to make us what we are.

—George Sarton, *Ancient Science and Modern Civilization*

By AD 500, the Christian Church had drawn most of the talented men of the age into its service, in either missionary, organizational, doctrinal, or purely contemplative activity.

—Edward Grant, *Physical Science in the Middle Ages*

IF YOU COULD look down on the Mediterranean world in the year 500, what would you see? An Ostrogothic king on the throne in Rome, doing his best to look like an emperor. An emperor in Constantinople, recreating the glories of imperial Rome on the shores of the Bosphorus, and, far to the south, in the very cradle of civilization, a Persian shahanshah, plotting his next move in the interminable war on his northern borders. A world of change, a world of confusion, a world where cities were shrinking, libraries were burning and little seemed certain anymore.

These conditions were not conducive to the preservation of texts or the pursuit of knowledge. Both need political stability, individual interest and consistent funding in order to flourish; all these were in short supply in AD 500. Even so, tiny pockets of learning endured and many books were kept safe. We have inherited great riches from our distant ancestors, but the reality is that huge swathes of ancient culture were lost on the long journey to the twenty-first century. Only a fraction has survived: seven of the eighty or so plays by Aeschylus, seven of the one hundred and twenty by Sophocles, eighteen out of ninety-two by Euripides. Many other writers have disappeared altogether, reduced to ghostly mentions in other works. In the late fifth century, a man called Stobaeus compiled a huge anthology of 1,430 poetry and prose quotations. Just 315 of them are from works that still exist—the rest are lost. Science fared a little better, but, still, important works like Galen's *On Demonstration*, Theophrastus' *On Mines* and Aristarchus' treatise on heliocentric theory (which might have changed the course of astronomy dramatically if it had survived) all slipped through the cracks of time. The texts that have survived, among them Euclid's *Elements*, Ptolemy's *Almagest* and the Galenic corpus, are the result of thousands of years of scholarship. The ideas they contain were filtered through the minds of generations of scribes and translators, transformed and extended by brilliant scholars in the Arab world who, in the late Middle Ages and the Renaissance, were gradually written out of history.

Attempts were sometimes made to recover books and, even in antiquity, people were aware of the danger that knowledge could simply disappear. According to Suetonius, the Emperor Domitian (AD 51–96) "went to a great deal of trouble and expense in restocking the burned-out libraries, hunting everywhere for lost volumes, and sending scribes to Alexandria to transcribe and emend them."[1] The only surviving manuscripts that were actually made in the ancient world (before around AD 500) are small fragments of papyri found on a rubbish tip in Egypt and some scrolls from the Villa of the Papyri at Herculaneum.* Everything else is a copy made at some point or other in the intervening centuries. Book production was a flourishing industry in the ancient world, with specialized markets and shops in towns and cities across the Mediterranean, so why do so few physical items survive? Until the fourth century, books were not books as we know them, but scrolls written on papyrus, which was made from reeds that grew in the Nile Delta. They were typically about three metres long, so, in order to read one, you had to unravel it at one end and gradually roll it back up again at the other, using special wooden sticks. The constant rolling and unrolling made the papyrus fragile and prone to tearing, so texts needed to be recopied onto new scrolls quite often. As we will see, by the time the more durable codex (made of parchment and wood) was

---

* These were preserved by volcanic ash after the eruption of Vesuvius in AD 79.

becoming dominant, the world had changed and not many people were making, selling or even reading books anymore. By AD 500, the Roman Empire had collapsed in Western Europe, and been radically reduced in the east. The vigorous culture of the ancient, pagan world was disappearing into the shadows of a new power: the Christian Church. For the next millennium, religion would dominate the world of books and learning in the West, while science found a new home in the Middle East.

The fifth century had been tumultuous, as the western half of the Roman Empire slipped away from imperial control into the hands of a collection of tribes from Northern Europe. Roman Hispania was now ruled by the Visigoths, with the northern part of the peninsula populated by the Alans and the Suebi. The Vandals had seized the northern strip of Africa, while Italy and even Rome itself had recently hosted the coronation (with full imperial splendour) of the Ostrogothic king, Theodoric. Meanwhile, the Franks were in the process of founding the country we now call France, and, across the Channel, armies of Anglo-Saxons were pushing deep into Britain. No longer held together by the power of Rome, societies in Western Europe began to turn inwards and become cut off from one another; cities shrank as people returned to the countryside and a simpler, more rural way of life. As the empire's system of travel and communication buckled, merchants could no longer transport their goods safely, so trade diminished drastically.

What was left of the empire, the eastern part, endured, but in a much-reduced form. The Emperor Anastasius (431–518), nicknamed Dicorus ("two-pupilled") because he had one black and one blue eye, ruled his dominions, which consisted of Asia Minor, Greece, the Balkans and parts of the Middle East, from his capital, Constantinople. In 500, the split between East and West was relatively new, and the social and cultural divisions that would characterize the following centuries had not yet taken hold. The imperial government in Constantinople still hoped to be able to recover at least part of the former Western Empire—in particular, Rome and its environs. This desire crystallized during the reign of the Emperor Justinian I (527–565), a powerful, energetic man, who scandalized the Byzantine elite by marrying his mistress, Theodora, twenty years his junior and a prostitute to boot.

Justinian's reign was long and eventful. He ordered an overhaul of the entire system of Roman law, began a huge programme of rebuilding in his capital (including the remodelling of the Hagia Sophia) and encouraged the production of silk, after two monks allegedly smuggled silkworm eggs and larvae back from China under their habits. With the help of his brilliant admiral, Belisarius, Justinian managed to win back parts of North Africa from the Vandals, got a foothold in Hispania and, most importantly of all, reconquered Sicily and most of Italy. The victory must have been sweet, but it was short-lived. The Ostrogoths did not give up their designs on Italy easily and Justinian found himself embroiled in a long and painful war on his western frontier, while the Persians attacked from the south and Turkic and Slavic tribes harried the northern border in the Balkans. Within a few decades of his death, all his territorial gains had been lost once again and the schism between East and West, its fault line running north–south between Greece and Italy, had begun to deepen.

Everyday life in late antiquity was extremely precarious, even for the wealthy 5 per cent or so of the population who weren't peasants or slaves. Disease and death stalked every household, hunger and disaster were never far away. Add to this hordes of invading barbarians trampling your crops and murdering your family and the picture becomes very bleak indeed. But there was one glimmer of light in the darkness, a faint spark of hope in the chaos—religion. The Roman Empire officially adopted Christianity in 380, and by 500 it had spread, in various forms, across Europe, the Middle East and northern Africa, replacing the wide range of cults, deities and faiths that fall under the umbrella term "paganism." Pagan faiths were varied and often localized; people believed in many gods, which were frequently closely linked to the natural world, and worship was geared towards trying to influence nature to ensure a good food supply and the health and happiness of the community. Christianity's insistence on the one true God created a stark choice—it was all or nothing—and eventually spelled the end for most of the old pagan faiths. As the Church gained power and popularity, its leaders became more and more determined to stamp out competing belief systems and Christianize the whole world. By AD 500, it was well on the way to achieving this mission.

It is striking that, at this point, a century before the arrival of Islam, there were many more Christians in the East than in the West, with monasteries and churches throughout Syria, Persia and Armenia. People clung to the promise of salvation. The idea that the more you suffered here on earth, the better your time would be in the afterlife was a potent shield against the desperate realities of everyday life in the fifth and sixth centuries. This doctrine was central to the success of Christianity's victory over paganism, which had traditionally championed the pursuit of happiness and denounced pain as evil. The triumph of suffering over pleasure had its most extreme expression in the early monasteries. Many were founded in this period; by 600, there were 300 in Gaul and Italy alone. In these often-isolated communities, the belief that, as the historian Stephen Greenblatt puts it, "redemption would only come through abasement"[2] dominated. Self-flagellation, self-deprivation and a lifestyle of acute asceticism was demanded of their inhabitants. But these monasteries were also places of peace and safety in a terrifying world, and, increasingly, the only place where anything approaching an education or a library could be found.

The battle between Christianity and paganism was long and

violent, and there were many casualties. Scholarship ended up in the no-man's-land between the two, as the prevailing force of the Church struggled to destroy or assimilate the philosophy, science and literature of the ancient world, which were, by their very nature, pagan. In 529, two crucial events tipped the balance even further in favour of Christianity. The Emperor Justinian closed the Academy in Athens, the centre of Neoplatonist philosophy and pagan resistance. The philosophers fled to Persia, taking their books and teaching with them, breaking the "golden chain"—the Athenian tradition of intellectual enquiry that stretched right back to Plato and Aristotle. Meanwhile, across the Ionian Sea, on the rocky hilltop of Montecassino in Southern Italy, a pious young Christian called Benedict founded a monastery and, with it, a new religious order that would spread across the world. In the centuries that followed, Montecassino became famous for its library and scriptorium, an important sanctuary for knowledge and education. As the doors of Plato's Academy closed for the last time, St. Benedict smashed the temple of Apollo that had endured for centuries and replaced it with a monastery. The symbolism couldn't have been clearer. A new era was shuddering into existence.

While it was true that Christianity had emphatically triumphed in the battle for people's souls, classical scholarship retained its hold over their minds. Everything about it was superior, from the brilliance of the ideas and sophistication of the arguments down to the beauty of the language and dexterity of the grammar—early Christian writings were notoriously clumsy, which was a matter of great embarrassment to churchmen. As one sixth-century writer put it: "We need a Christian and a Pagan schooling; from the one we gain profit for the soul, from the other we learn the witchery of words."[3] But it was one thing recognizing the value of a classical education—quite another thing protecting the schools that provided it from the turbulence of a changing world. Some schools survived the Ostrogothic invasion of Italy in the fifth century, and Justinian was keen to cement his reconquest of Rome by re-establishing higher education in the city. The scholar Cassiodorus (*c*.485–*c*.585) dreamed of founding a theological university there, but these plans came to nothing. The Lombard invasion in 568 sounded the death knell for traditional education in Italy, which had, at any rate, only ever been available to a small minority of wealthy male children. The privileged few who could afford it educated their children at home, but monasteries increasingly had a monopoly on teaching, with the inevitable emphasis on Christian literature and doctrine.

It was much the same story for book production, which dwindled across the Mediterranean during the fourth and fifth centuries. Some commercial book production continued in big cities like Rome, but on a much smaller scale than previously. The majority of books were copied privately by individuals who had access to the texts they wanted through friends or networks of scholars. By the year 500, secular book production had effectively gone underground; in contrast, the output of monastic scriptoria grew dramatically as entire new genres of religious literature were created, like hagiography (the stories of saints' lives). Cassiodorus, unable to found his university in Rome, went instead to his family estates at Squillace, on the southern coast of Italy, and founded a monastery, Vivarium, inspired by the school at Nisbis, in Syria, he had heard about and possibly visited while living in Constantinople. A devout Christian, Cassiodorus was also a passionate believer in the classical curriculum, which was organized into the trivium (rhetoric, logic, grammar) followed by the quadrivium (arithmetic, geometry, astronomy and music). He filled the library at Vivarium with texts on these subjects and transformed the production of manuscripts in his scriptorium by developing proper standards and methods for copying. As one of the few notable scholars of his period, Cassiodorus played a vital role in the survival of classical culture in Italy, saving books from the smoking ruins of Roman libraries, preserving and reproducing them, ensuring that they reached the next and future generations—they went on to form the framework of the medieval educational system. Having spent twenty years living in Constantinople, he was also one of the last scholars to bridge the growing chasm between East and West, and to bring the Greek culture and language of Byzantium back to Italy in the form of several Greek manuscripts, which were housed in a special cupboard in the library at Vivarium.

In 523, Cassiodorus was appointed magister officiorum (chief advisor) to the Ostrogothic King of Italy, Theodoric, taking over from the only other major scholar in Italy at that time, Ancinius Manlius Severinus Boethius (480–524). Boethius had gone much further than Cassiodorus in his promotion of ancient scholarship. For Cassiodorus, it was always the handmaiden of Christianity, to be studied only with the ultimate aim of getting closer to God. Boethius, on the other hand, believed in the value of secular learning for its own sake, and he had embarked upon an ambitious project to translate all the Greek texts necessary for the study of the classical curriculum. His efforts were cut short when he was imprisoned and later executed for his suspected role in a conspiracy against King Theodoric. Had all his translations been preserved and passed down, the story of the transmission of ancient science could have been very different. As it is, we only have vague clues about what he actually did translate, but this seems to have included part of *The Elements* and some of Ptolemy's writings (but not *The Almagest*). There are various mentions of a Latin translation of (at least parts of) *The Elements*, by Boethius, and ghostly fragments of it can just be made out in a fifth-century palimpsest in the Biblioteca Capitolare in Verona, which features material from Books 1–4, but without the diagrams and proofs, so it would have been of limited use. There were probably very few copies of it, and those that existed were neglected. By the ninth century, only shreds remained. We know little about this version of Euclid's great work, but, at the very least, it alerted scholars to the existence of a much deeper source of knowledge on the subject of mathematics.

Raphael painted the figures in *The School of Athens* reading or holding books, when in fact they would have written on papyrus scrolls. The codex, or book, did not come into use much before the fifth century and was made of parchment—treated animal skins— rather than papyrus, which was made from reeds; paper mills arrived in Europe in the eleventh century, although there were several in the Islamic world centuries before that.* At best, papyrus only lasts for a couple of hundred years before the text needs to be recopied onto a new scroll. Parchment lasts longer, but only if it is kept in the right conditions, free from damp, rodents, worms, moths, fire and a host of other potential enemies. The codex was initially a Christian phenomenon, and grew in popularity between the fourth and eighth centuries. If we narrow the process of transmission down to a single, hypothetical strand, it is feasible that Ptolemy originally wrote *The Almagest* on a papyrus scroll in second-century Alexandria. That scroll would have had to be recopied at least twice for it to survive until the sixth century, at which point it might well have been copied onto parchment and bound into a book. This, too, would need to be recopied every few hundred years to ensure that it survived (again assuming that it escaped the usual pests, damage and disasters) and was available to scholars in 1500. It is therefore likely that *The Almagest* had to be recopied at the very least five times during the period 150–1500. The question is, who copied it, and where did they find it?

The fate of every text was determined by what was happening outside the walls of the library or private house in which it was

---

* Paper was also imported to Europe during the Middle Ages, often from Damascus, hence it was known as "charta damascena." It was expensive, but, as production developed in Europe, the price fell and it gradually replaced parchment.

shelved. In the tumultuous years of late antiquity, the tectonic plates of political, social and religious life were shifting and rearranging themselves in profound ways. The world of scholarship gradually moved from the public, secular sphere into the hushed cloisters of monasticism. This movement was visible in other areas of life, too: the topography of cities began to change as the Church moved in to fill the void left by the "res publica"—the Roman state. Power melted away from the state into the hands of private individuals and religious leaders. Huge churches rose up in the ancient forums, temples were destroyed or converted, the public space of the city was being Christianized and the bishops moved to centre stage. Like schools, public libraries were another casualty of this process. With no one paying for their upkeep, they fell into disuse and decayed. Anyone interested in subjects like maths and astronomy had to pursue them privately, so the fragile networks of scholars shrank even further.

With medicine, the story was slightly different because of the constant and urgent need for it. Medical knowledge was always useful, always relevant, so books on medicine were constantly in demand, and would have been available in the majority of libraries in late antiquity. Medicine was always a multilayered enterprise, with basic care being carried out by people within their own homes, while the next level was the local healer, or wise man or woman, who would have particular knowledge of local plants and herbal remedies. This knowledge was oral, however, and its practitioners, for the most part, were illiterate. Educated doctors were few and far between; their training varied enormously and they mainly served wealthy urban clients. Religion had also played an important role in ancient medicine; centres of medical teaching at Smyrna, Corinth, Cos and Pergamon had healing shrines that attracted supplicants in the same way that Catholic shrines do today. The doctors who worked in them would have treated patients and trained medical students using books that they had collected. But, as centres of pagan faiths, many were destroyed when Christianity gained the ascendancy.

The shrine dedicated to Asclepius, the Greek god of healing, in Pergamon, in Anatolia, attracted thousands of supplicants and became a famous centre of medical learning. Galen was born and educated there before he set off for Alexandria and then Rome, and the city was also home to an important library—200,000 books, according to the historian Plutarch. Founded in the third century BC by the Attalid dynasty, the Roman writer Strabo (64 BC–AD 24) mentions it when he discusses what happened to Aristotle's books: "But when they [Aristotle's heirs] heard how zealously the Attalid kings to whom the city was subject were searching for books to build up the library in Pergamon, they hid the books underground in a kind of trench."[4] Unsurprisingly, the books didn't benefit from being buried in a trench, where they were "damaged by moisture and moths"— much better that they had ended up on the shelves of the Pergamon library, with its walls specially designed to allow air circulation and prevent damp.

Pergamon's greatest rival as an intellectual centre was the city of Ephesus. Towards the end of the second century, Ephesus began to pull ahead in the race to be "first of Asia."[5] Pergamon's decline during the third century was hastened by earthquakes and Gothic raids. Christianity arrived in the city, heralding the building of many churches. But, though the Pergamese then enjoyed a period of stability, it was short-lived. In the following century, the population shrank as the persecution of non-Christians increased (several pagan cults lingered in the city) and plague decimated those who remained. In the meantime, Ephesus was enjoying its heyday. Capital of the Roman province of Asia, it was a thriving port city, famous for its Temple of Artemis—one of the wonders of the ancient world. A marble colonnaded avenue led visitors from the harbour up through the city, past shops selling souvenirs of Artemis, to the magnificent

1. The reconstructed facade of the Library of Celsus in the ruined city of Ephesus, built in the second century as both a mausoleum to a Roman senator and a repository of some 12,000 scrolls. These were stored in cupboards placed in niches with double walls specially designed to control levels of humidity and temperature.

amphitheatre that could accommodate 24,000 people. In AD 117, a great library was built there in honour of the Roman senator Celsus, who was buried in a mausoleum beneath it. This impressive building housed 12,000 scrolls, making it the third largest collection, after those of Alexandria and Pergamon. The inside was damaged by the Goths when they attacked the city in 268, but the grand facade remained standing until an earthquake finally knocked it down in the tenth century.

Ephesus was an early centre of Christianity, too—St. Paul lived there in the mid first century, while St. John spent his last years there, writing his gospel. As the new religion took hold, the old pagan sanctuaries inevitably suffered. The Temple of Artemis was first vandalized and then abandoned, its statues buried deep underground, where the demons that inhabited them could not threaten the Christian citizens above. The city's other temples were destroyed or converted into churches. No doubt many texts perished at the same time. As the river mouth gradually filled with silt, a new alluvial plain was formed and the coastline changed dramatically. Ephesus was cut off from trade and communication (today it is several miles inland); by the thirteenth century, it was all but deserted.

So what happened to all the scrolls in the libraries of Ephesus and Pergamon? One legend claims that the Roman general Mark Anthony took them from the library at Pergamon and gave them to his lover, Cleopatra, for the library at Alexandria. Did local scholars try to save some of them? Were they taken to safety and painstakingly recopied and preserved, handed down through generations of families, or hidden in the ruins of ancient temples? Some of them must have been, because, as we will discover, Anatolia was a major focus of the Abbasids' search for ancient Greek texts in the ninth century. A tenth-century Arabic source describes an ancient temple, about three days' journey from Constantinople, which "had been locked since the time when the Byzantines became Christians." The Arabs persuaded the Byzantine official to open the gates, "and, behold, this building was made of marble and great coloured stones," and, inside, "there were numerous camel-loads of ancient books."[6]

But first we need to go back in time, to the very beginning, when Euclid, Ptolemy and Galen sat down to write their books, to see where the first copies were made and disseminated. Galen mainly lived and worked in Rome and Pergamon, but Ptolemy and Euclid both wrote their masterworks in the city that was the intellectual heart of the ancient world: Alexandria—home to the library that has inspired and overshadowed libraries ever since.

TWO

Alexandria

The advantages of the city's site are various; for, first the place is washed by two seas, on the north by the Aegyptian Sea, as it is called, and on the south by Lake Mareia, also called Mareotis . . . The city as a whole is intersected by streets practicable for horse-riding and chariot-driving, and by two that are very broad, extending to more than a plethrum in breadth, which cut one another into two sections and at right angles. And the city contains most beautiful public precincts and also the royal palaces, which constitute one-fourth or even one-third of the whole circuit of the city; for just as each of the kings, from love of splendour, was wont to add some adornment to the public monuments, so also he would invest himself at his own expense with a residence, in addition to those already built, so that now, to quote the words of the poet, "there is building upon building." All, however, are connected with one another and the harbour. The Museum is also part of the royal palaces . . .

—Strabo, *Geography*

When Demetrius of Phalerum was put in charge of the king's library he was lavished with resources with a view to collecting, if possible, all the books in the world; and by making purchases and copies he carried out the king's intention as far as he could.

—*Letter of Aristeas to Philocrates*

THE GREAT LIBRARY of Alexandria, founded around 300 BC by the Egyptian king Ptolemy I, has always been the ultimate symbol of scholarly endeavour. It was here that the idea of encompassing knowledge in one place by collecting a copy of every single text was born. This "dream of universality" has haunted book collectors and librarians ever since, and lies at the heart of modern copyright libraries, which are entitled to one copy of each book published in their own country. Like all the most successful bibliophiles, the kings of Egypt and their librarians were doggedly unscrupulous in the pursuit of this dream: stealing, borrowing, begging—anything to increase the collections. They ordered that all ships passing through Alexandria should be searched and any scrolls found on board confiscated. These were then labelled "from the ships" and shelved in the Library. When the Athenians lent valuable scrolls for copying, the Egyptians refused to return them, choosing instead to keep the originals and send back copies, forfeiting the huge sum of money they had paid as surety. This aggressive acquisition policy paid off and within a couple of decades the Library contained thousands on every subject from cookery to Jewish theology—a collection unequalled, both in size and subject matter, anywhere on the planet. But the Ptolemaic kings did not just collect books, they collected minds as well. They established a community of scholars in the shrine they had built to glorify the Muses—the nine Greek goddesses who inspired the arts and sciences. It became known as the Museum (*Mouseion*) and was closely linked with the Library; scholars from across the Mediterranean world were invited to come and work there. As time went on, a daughter library was created in the Temple of Serapis (the *Serapeum*) to house the ever-increasing collections.

The question of where the idea for the Library of Alexandria originated has long puzzled historians. Aristotle is the first person known to have collected books privately, and writers from Strabo (64 BC–AD 24) onwards have suggested that he inspired the foundation of libraries in many of the cities his pupil Alexander the Great conquered and created. The idea of a universal collection also probably came from Aristotle. His intellectual interests were informed by a similarly comprehensive scope and another of his pupils, Demetrius of Phalerum, was instrumental in the design and creation of the Library of Alexandria. The city was founded by Alexander when he conquered Egypt in 331 BC. According to legend, he personally selected the site, conveniently located in the Nile Delta between Lake Mareotis and the sea, with excellent transport routes and two large natural harbours on the Mediterranean coast. When Alexander died, Egypt, by far the wealthiest part of the vast Greek Empire, passed to one of his most trusted generals, Ptolemy Soter. The remainder was divided between two other generals, and together the three areas were known as the Hellenistic kingdoms. Soter proclaimed himself king and founded a dynasty which went on to rule Egypt for 275 years, ending only with the dramatic suicide of Queen Cleopatra. This longevity was certainly not a given. Soter was an upstart Macedonian nobleman; it took a massive programme of political, social, military and cultural development to establish his position as the unchallenged ruler of Egypt. Competition with Alexander's other heirs was a constant preoccupation for both Soter and his son, Soter II, and while some of this took place on the battlefield, a significant proportion was played out on the desks and bookshelves of the Library and the Museum.

As the beautiful new city spread out across its harmonious grid,

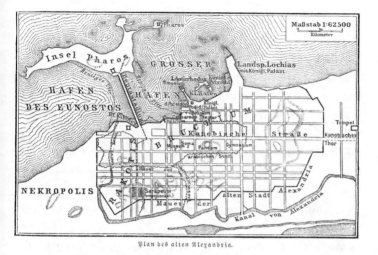

2. A nineteenth-century German map of ancient Alexandria showing Pharos Island, the harbours, the Royal Quarter (where the Library and the Museum were located), the Jewish Quarter and the city's grid street system. Lake Mareotis is at the bottom, and the Serapeum is just above it on the left of the map.

a new cultural identity was forged that assimilated the traditions of ancient Egypt with that of the Hellenistic world. Initially, this involved subjugating native Egyptians and their culture, while surrounding the Ptolemies (who were of course Macedonian) with a much-needed aura of Greek (and, in particular, Athenian) legitimacy and emphasizing their connection with Alexander the Great.[*] This is clearly visible in the Museum, which, like the Library, was also inspired by Aristotle and his Lyceum in Athens. Both institutions were situated within the shrine to the Muses, and both were communal enterprises. The Ptolemies generously endowed their Museum, paying the scholars well, exempting them from tax and providing them with board and lodging in a special part of the pal-

---

[*] When Alexander died in Babylon, Ptolemy Soter seized his body and had it brought back to Egypt to consolidate his position as Alexander's main heir.

ace complex. Around the time of the birth of Christ, the Roman geographer Strabo came to visit Alexandria and described the Museum thus: "it has a covered walkway, a hall with seats (exedra) and a large house, in which there is a common dining hall for the learned men who share the Museum." These scholars "have communal possessions and a priest in charge of the Museum, who used to be appointed by the kings, but is now appointed by Caesar."[1] With this kind of support on offer, it is hardly surprising that so many intellectuals made the city their home. If you were a scholar, there was no better place to be.

Thanks to the Ptolemaic kings, Alexandria became the most important centre of learning in the ancient world, seizing the crown of Greek cultural dominance from Athens and projecting a new vision for state-sponsored scholarship that was admired and emulated across the Mediterranean. While the scholars argued "interminably in the chicken coop of the Muses"[2] and the bookshelves of the Library filled with scrolls, the city was growing. Baths, brothels, houses, shops and shrines were built along the broad perpendicular streets, as communities of different nations—Egyptians, Jews, Greeks and, later, Romans—settled and worked, lived and died. Soon Alexandria was one of the largest cities on earth, "the unrivalled centre of world trade"[3] that exported huge quantities of grain, papyrus and linen, produced on the fertile Nile plains, shipped down the river to the city and then sent off for sale across the Hellenistic world. As the gatekeepers to the Mediterranean for merchants from Africa, Arabia and the East, Alexandrians took a healthy cut of the lucrative trade in gold, elephants, spices and perfumes, which were shipped in from the south and east across Lake Mareotis. The great Pharos lighthouse, 120 metres tall and another of the seven wonders of the ancient world, towered over the harbour, a symbol of Alexandria's brilliance, beaming out across the sea.

Alexandria lay at the centre of a large network of cities, among them Athens, Pergamon, Rhodes, Antioch and Ephesus, and, later, Rome and Constantinople. Books and scholars moved freely between them in the thriving marketplace of ideas. Clever young men from across the Hellenistic world were educated in their home towns, before setting off in search of better teachers, bigger libraries and higher knowledge. Books for primary education would have been available to them at school or in the local public library—of which there were a surprisingly large number in the ancient world. Most towns had a collection, but only the large libraries in cities would have contained scientific texts in any number—most copies of the books we are following here would have been owned privately, by specialist scholars. Unlike literature, with its hundreds of poems, speeches and plays copied, sold and read all over the Mediterranean, science made up a tiny proportion of ancient writing and was only of interest to an educated elite—just 144 mathematicians are known in the whole of antiquity. While the great libraries fill history books, it is the small private collections, carefully shelved behind closed doors, that were crucial to the transmission of science. No scholar of mathematics, medicine or astronomy would have been able to study without owning a few books of their own. Nor would they have been able to teach the students who gathered around them. Because these kinds of collections were private, there is very little historical evidence for their existence, but we can safely assume that they would have been accumulated throughout a scholar's career, beginning at school. Scholars would have borrowed texts from their teachers and colleagues and made copies for themselves, or got their slaves or students to do it for them.

Collaboration was vital—scholars had to band together to share their resources, and they tended to do this in large cities, where there was already a tradition of learning and a library—it was very difficult to make any headway in science in isolation. This is why places like Alexandria played such an integral role in the history of science.

Anyone interested in academic learning knew that, if they were to make progress, get hold of texts and have the chance of working with other scholars, they would have to travel to one of these centres. The chances were that they would be pointed in the direction of Athens or Alexandria by their teachers, who had likely studied in these places in their youth. In an age where knowledge and ideas were extremely hard to access, networks of like-minded people underpinned intellectual enquiry, but they were very small. Archimedes, the most brilliant scientist of the ancient world, lived in Syracuse in Sicily—a relative backwater when it came to scholarship. When his collaborator, Conon, died, Archimedes was left desperately casting around for someone "versed in geometry" to replace him. He also complained in the introduction to his treatise, *Spiral Lines*, that, "Though many years have elapsed . . . I do not find that any one of the problems has been stirred by a single person."[4] These plaintive cries show how few people studied science at this level. The small number who did had to work together and share their expertise and resources, especially books.

Alexandria was the capital of the intellectual world for over a millennium, so it is no coincidence that the three men whose ideas we will follow in this book all lived and studied there. In the early decades after the city was founded, Ptolemy I actively sought out scholars to come and help him transform his city into a place of learning to rival Antioch, Athens and Rhodes. The evidence is scarce, but it seems that Euclid was one of these scholars, and that he came from Athens around 300 BC, where, just a few decades before, Plato had been busy teaching mathematics and philosophy in the Academy, under the sign proclaiming, "Let no one ignorant of geometry enter here." Euclid would certainly have brought books with him to Alexandria, which would have been copied and added to the Library. Euclid settled into his new home, where he was supported by Ptolemy I, and set to work with other like-minded scholars, probably in the Library itself. The fragments of information about his personality, which may or may not be genuine, paint him as a conscientious, hard-working man, "well disposed toward all who were able in any measure to advance mathematics . . . and although an exact scholar not vaunting himself."[5] The huge amount of work and organization that must have gone into producing *The Elements*—to say nothing of his other works—supports this view. A bookish, serious man, who loved maths passionately, Euclid stayed in Alexandria, and the school of mathematics that formed around him continued for centuries. His journey south across the sea, away from Athens, took the study of mathematics out of the shadow of philosophy, allowing it to become a subject in its own right.

Euclid was not antiquity's most brilliantly original mathematician—that accolade is universally awarded to Archimedes—but he did write the greatest mathematical textbook of all time. In *The Elements*, he gave the world a masterly explanation of the universal principles of mathematics, presented in such an ordered, lucid way that it was still being used as a textbook 2,300 years later and, according to one scholar, "has exercised an influence on the human mind greater than that of any other work except the Bible."[6] *The Elements* is a systematic survey of the mathematical knowledge available in the early third century BC, so Euclid stands at a crucial point in the history of mathematics, at the end of an ancient tradition that stretches back at least 2,000 years, and at the beginning of the epoch of which we are the inheritors. *The Elements* ushered in a new era of mathematics, standardizing the fundamental ideas of the subject and elevating it, from merely solving specific, localized problems, to a set of principles that could be universally applied, and universally proved—something that could be practised and enjoyed for its own sake.

In order to achieve this, Euclid must have had access to a large number of mathematical texts—those he owned personally, supple-

3. Pages from the manuscript of Euclid's *Elements* in Greek, which was written on parchment in Constantinople in 888 before being bought by Arethas of Patrae, whose annotations can be seen in the margins and underneath the text. It is considered to be the oldest complete copy of the text and is also the oldest dated book by a classical Greek author.

mented by others already in the collections at Alexandria. Given the volume of material he covered, it is likely that he had the assistance of a group of scholars working under his direction. Having methodically assessed the information available to him, Euclid began by setting out the absolute basics, starting with definitions of the fundamentals—"a point is that which has no part," "a line is a breadth-less length."[7] Then he presented each topic logically, one at a time, organizing everything so that it made sense and each section led naturally on from the last.

The Elements is divided into thirteen books. The first focuses on Pythagoras' theorem, Book 2 introduces geometric algebra, Books 3 and 4 are about circles, Book 5, the most admired, looks at proportion, while Book 6 applies it to geometrical figures. Books 7, 8 and 9 are concerned with numbers, Book 10 square roots and Books 11 to 13 explain solid geometrical shapes. Euclid was not the first person to try to systematize mathematical knowledge, but his version was so brilliant, so much clearer than anything that had gone before, that it quickly became the standard text on mathematics. The disadvantage of this was that the scribes and scholars no longer copied out the older works on which it was based. *The Elements* eclipsed and replaced them so profoundly that only one mathematical treatise has survived that pre-dates it. Euclid transformed his subject by creating universal standards and methods for practising mathematics—introducing the demonstrative method, an idea he probably got from Aristotle, that has been used not only in maths, but in all the exact sciences ever since. He explains the theories by a series of definitions, called axioms (from the Greek, meaning "things we can take for granted"), using a limited and rigorously defined terminology so that everyone can understand what he means; he then demonstrates them using diagrams and geometrical proofs, labelled with letters of the alphabet—a scientific practice that hasn't changed for over 2,000 years.

We do not know how *The Elements* was received by Euclid's peers, nor how many copies were made early on, but we can assume that at least one was produced for the Library in Alexandria, where it could be consulted and recopied by other scholars. It is safe also to assume that copies were sent to the other major intellectual centres of the ancient world—Athens, Antioch and Rhodes—to enhance their mathematical collections. The early history of this seminal book is patchy; there are only small traces of its existence in the first few centuries after Euclid died. Second century BC shards of pottery with shapes and workings from Book 13 scrawled on them were found on Elephantine Island (now part of the modern city of Aswan), so someone in a remote part of Egypt was working out Euclid's ideas, and not just the basic geometry from the early sections of *The Elements*, but the final, most complex book, which was the culmination of the whole project. Papyrus fragments of Euclidean diagrams also turned up in an ancient rubbish tip near Oxyrhynchus in central Egypt, along with small pieces of thousands of other manuscripts and documents—preserved by the arid climate in the desert sands. The Oxyrhynchus fragments, written between AD 75 and 125, are the earliest and most complete examples of Euclid's diagrams. These finds show that *The Elements* was definitely being read and used, and therefore recopied and preserved, in the period after Euclid's death, but it is difficult to draw general conclusions about its popularity from so little evidence.

In the first century BC, the vibrant tradition of commentaries explaining *The Elements* began, with the astronomer Geminus, who lived in Rhodes, providing definite evidence that at least one copy of Euclid's masterpiece had made its way there. As the different branches of science developed, scholars increasingly took up the work of previous generations and wrote detailed commentaries explaining and clarifying the original text, often in columns beside it, but sometimes in separate books. Commentaries would become one of the most common forms of scientific writing, and, as "the dominant cultural vehicle"[8] in late antiquity, played a vital role in the transmission of ideas from generation to generation. Six mathematicians wrote major commentaries on *The Elements* in the period 300 BC–AD 600, demonstrating a small yet consistent level of interest. In the earlier Hellenistic period, mathematical study was characterized by originality and discovery; in contrast, these works are indicative of the systematic nature of mathematics after Euclid, a period of assimilation and organization rather than innovation.

The most influential commentary of all was written by Theon of Alexandria (AD 335–405), another famous mathematician, and father of the great philosopher and astronomer Hypatia.[*] When Theon came to read *The Elements*, it was 600 years old and in need of modernization. He edited and clarified Euclid's work, adding new proofs, adapting the language and even removing sections that didn't seem to make sense. His new edition was very successful; it was recopied many times and spread throughout the Mediterranean world. It became the standard version, the single master source for all other editions of the text throughout the Middle Ages and beyond, right up to 1808, when something astonishing happened. A French scholar called François Peyrard was sorting through a pile of books that Napoleon had "obtained" from the Vatican Library and taken back to Paris. Among them was a manuscript of *The Elements* that was very different from the Theonine editions. Scholars soon realized that this copy of the text did not contain Theon's edits and additions—it was an older and therefore a more pristine version—closer to Euclid's original text. The manuscript Peyrard found had been copied in Constantinople in around AD 850, so it had lain hidden for almost a thousand years, eluding centuries of scholars, and it represented an exciting new thread connecting us back to Euclid himself. Eighty years later, J. L. Heiberg, a Danish Professor of Philology, used the manuscript, along with other editions, fragments and references, to produce a definitive version of the text. The Heiberg edition is still the basis of the modern standard edition of Euclid's *Elements*.

In the centuries after Euclid's death in around 265 BC, intellectual life in Alexandria continued to flourish, especially in the sci-

---

[*] Hypatia's story is one of the most tragic and compelling in the whole of antiquity and has made her the best-known female scientist of the period. A leading figure in the Alexandrian intellectual scene, she was educated by her father Theon and worked alongside him, but became the focus of Christian antagonism towards pagan culture and was murdered by a fanatical mob.

ences, literature and medicine. Now that the Ptolemaic dynasty was secure, the Greek elite began to take an interest in the riches of ancient Egyptian culture. They adopted some of the local customs (including the questionable tradition of sibling marriage), Egyptian texts were translated into Greek in the Library and intellectual traditions were merged. There was also a groundbreaking programme of Jewish translation, with the first Greek version of the Pentateuch (known as the Septuagint) produced from the Hebrew by a specially selected group of Jewish elders.

As anyone who collects books knows, you don't need to have many on your shelves before some kind of organization is necessary. The early librarians of Alexandria quickly realized that they needed to keep a record of the collections, and they needed to shelve them in some kind of order so that readers could easily find specific titles. Callimachus of Cyrene, an accomplished poet who was associated with the Library in the mid third century BC, produced a detailed scroll catalogue, called the *Pinakes*. Only fragments of the original 120 volumes survive, but they reveal that the texts were divided into the following categories: rhetoric, law, epic, tragedy, comedy, lyric poetry, history, medicine, mathematics, natural science and miscellanies. This was the first serious attempt to organize knowledge into a universal scheme and, as such, it was a turning point in the history of ideas. The *Pinakes* also gave a potted bibliography of each writer and listed their books, establishing not only a canon, but also the tradition of producing material describing both the author and the work itself. This meta-textual tradition reached its highest expression in the hallowed arena of Greek literature, with hundreds of scholars studying, editing and arguing over the plays and poetry of, among others, Homer, Euripides and Sophocles, whose works were copied and sold across the Greek-speaking world—and many of these copies are the basis for the editions that have come down to us today.

The Ptolemies personally led this intellectual odyssey. The first four kings of the dynasty were known for their interest in various scholarly pursuits—one was a poet, another was fascinated by zoology. All Ptolemaic monarchs, right down to Cleopatra (who was herself a linguist), attended events and debates at the Museum, and intellectual engagement was a salient feature of their rule. However, in the first century BC, a new world power was on the rise, and before long its armies arrived to conquer the dazzling city. By 80 BC, Alexandria was formally under Roman rule, but its intellectual life was allowed to continue, unimpeded. Unfortunately, the Library had suffered its first major loss in 48 BC, when Julius Caesar attacked the city and his troops burned down a huge storehouse of scrolls in the harbour. This was no doubt unintentional; after all, Caesar was known for his love of books and had been responsible for introducing the public library to Italy (like so much else in Roman culture, it was an idea borrowed from Greece). And despite the lost scrolls, and despite Roman occupation, a period of great prosperity followed as Egypt supplied her new masters with grain, and the city continued to be the main centre of Greek learning.

At the end of the first century AD, a young man named Claudius Ptolemy was one of the thousands of people living in this great metropolis. The combination of his Graeco-Roman first name, Claudius, and his Egyptian surname, Ptolemy (no relation of the royal dynasty, as many later scholars believed), shows how intertwined the two cultures had become in Alexandria. Claudius Ptolemy has left us frustratingly little information about his life, but he was probably educated in Alexandria and spent time studying in the Museum to build up a body of knowledge on which to base his own work. What we do know is that he was captivated by astronomy, that he spent his nights gazing up at the stars, dedicating his life to trying to capture and make sense of their movements; doing so, he believed, brought him closer to the divine.

Ptolemy must have been an extremely curious man, someone

who was fascinated by the world in which he lived and determined to enrich our understanding of it. He wrote many books, covering a bewildering array of subjects: astronomy, mathematics, geography and astrology, but also musical theory and optics—the study of light and vision. For many centuries after his death, he was best known for his *Geographia*, a radical attempt to describe and map the known world. Today, he is most famous for writing the *Mathematical Syntaxis*, a description of the heavens and celestial bodies, which became known as *The Greatest Compilation* (translated into Arabic as *Al-Majisti* and, from there, Latinized into *The Almagest*). Gerd Grasshoff, author of *The History of Ptolemy's Star Catalogue*, explains its extraordinary influence: "Ptolemy's *Almagest* shares with Euclid's *Elements* the glory of being the scientific text longest in use. From its conception in the second century up to the late Renaissance, this work determined astronomy as a science."[9]

Like Euclid, Ptolemy worked in the Library of Alexandria, probably alongside other scholars, sorting through all the astronomical works he could find, from Babylonian, Egyptian and Greek traditions, among others. He assessed and tested theories and observations, before setting out the information in a clear, rational way and adding his own original contributions. As he explains in the preface to *The Almagest*:

> We shall try to note down everything which we think we have
> discovered up to the present time; we shall do this as concisely as
> possible and in a manner which can be followed by those who
> have already made some progress in the field. For the sake of
> completeness in our treatment we shall set out everything that
> is useful for the theory of the heavens in the proper order, but
> to avoid undue length we shall merely recount what has been
> adequately established by the ancients. However, those topics
> which have not been dealt with by our predecessors at all, or not

as usefully as they might have been, will be discussed at length to the best of our ability.[10]

Ptolemy's approach to the universe was mathematical, an alternative to Aristotle's physical description of the heavens, which posited the idea that the stars were arranged on revolving crystal spheres. In this, Ptolemy was guided by *The Elements* and its crucial role in the development of mathematics; he not only used Euclid's great work as a stylistic model, but also explicitly says that he is counting on the reader having a firm grasp of geometrical theory. His models of planetary movement were created using Euclidean geometry and were explained using the same system of postulates and diagrams. And, like *The Elements*, Ptolemy's work is divided into thirteen books that take the reader on a tour of the night sky, beginning in Books 1 and 2 with the mathematical knowledge needed, then focusing on the sun and moon in the following three books. Book 6 is all about eclipses, while Books 7 and 8 catalogue the stars. The final five books are about the planets, and represent Ptolemy's most important and original contribution to astronomy; they describe a complex mathematical model showing how the planets move, based on data he took from the earlier Greek astronomer Hipparchus,[11] and from his own observations. Ptolemy's model of the heavens was geocentric, with the earth fixed at the centre, and, as we will see, it would remain the dominant ideology until 1543, when Copernicus introduced the notion of a heliocentric universe, with the sun at the centre.

There are many similarities between Euclid and Ptolemy, so it's easy to forget that they were separated by four centuries. Alexandria was a completely different city by the time Ptolemy walked its streets, but the tradition of learning that began in Euclid's lifetime had endured and it was still a thrilling place for anyone with intellectual interests. The Library attracted students and scholars from across the Mediterranean, and, in turn, they added their own works and books they brought with them to the collections. Scholarship, then as now, thrives on collaboration and shared ideas, the passing on of knowledge. Ptolemy was able to write *The Almagest* because Alexandria provided him with the conditions he needed to produce such a masterpiece, conditions that simply weren't available anywhere else at that time.

Ptolemy based his work on observations of the heavens made by earlier astronomers, in particular ancient Babylonian sources and data produced by Hipparchus, but he also made observations of his own over the period 26 March 127 to 2 February 141. At this point in the history of astronomy, such observations simply involved staring up at the night sky and making notes on the positions of the stars and the planets. Ptolemy used various instruments to take measurements, including rulers, armillary spheres and astrolabes, but they were not at all accurate. Invented sometime in the second century BC, astrolabes were complex devices, circular plates of brass with intricate projections of the celestial sphere carved onto them, that could measure angles and help to predict the peregrinations of the stars. Throughout history, the necessity of designing effective instruments has been one of the greatest challenges facing astronomers, and is at the heart of their struggle to produce the most accurate data on which to base their scientific observations.

While Ptolemy was busy systematizing astronomical theory in Alexandria, another young man arrived in the city to pursue another kind of knowledge—medicine. In contrast to the shadowy figures of Euclid and Ptolemy, Claudius Galenus leaps out at us from the pages of history, many of which he wrote himself. Galen, as he came to be known, was one of the most prolific writers of antiquity, a prodigious self-publicist (he repeatedly claimed that he had "brought

4. A reconstruction (with original frieze panels) of the monumental Altar of Zeus, just one of the extraordinary public buildings in ancient Pergamon, where Galen grew up.

medicine to perfection"),[12] whose genius as a physician took him from his home in Pergamon to the bedside of the Emperor in Rome. We can imagine him arriving by boat in Alexandria, a clever, energetic young man, filled with the arrogance of youth, determined to make his mark on the world.

Galen was fortunate enough to be born in Pergamon in AD 129. At this point, the city was enjoying a period of extraordinary success: it was protected and favoured by the Roman state, and huge revenues from farming, minerals and trade were flowing into the municipal coffers, and into the pockets of Pergamon's citizens. Galen's father was a wealthy and important architect—an excellent career choice in a city that was undergoing an impressive programme of rebuilding and revitalization. A large community of specialist workmen and masons lived and worked there—the streets would have rung with the sound of their hammers as they hewed new temples, lecture halls and theatres from rock and marble. These new constructions augmented a city that was already one of the most magnificent in the ancient world, home to a vast altar dedicated to Zeus, a vertiginous amphitheatre cut into the mountainside and an acropolis that was modelled on the one in Athens.*

Galen received an outstanding education in the lively intellectual atmosphere of the gymnasia and library in Pergamon, under the watchful eye of the statue of Athena, grey-eyed goddess of wisdom. He may have been on track to follow in his father's footsteps, but, when he was seventeen years old, something happened that changed the course of his life. Galen's father had a dream in which the god Asclepius told him that his son should be a doctor. From that moment on, Galen focused on medicine. It was also the beginning of his own profound personal relationship with the god whose advice, imparted via dreams, he followed for the rest of his life. Galen trained under the masters who worked in the Asclepeion (the Temple of Asclepius)—part hospital, part spa and part shrine, and one of the most important centres of healing in the ancient world. People came from far and wide to be cured there and the little offerings (votives) they left to appease or thank the gods are poignant evidence of mankind's desperation in the face of illness. His father died when Galen was twenty, and soon afterwards he set off for the medical school at Smyrna, before moving on to Corinth. From there, he travelled to Alexandria—at that time, a great centre of medicine and the only place you could study human skeletons—a vital destination for an ambitious young doctor.

Galen was far from complimentary about Alexandria, but, in spite of the fact that he complained about everything, from the food to the climate and the Egyptians themselves, he stayed in the city for five years and learned a lot about anatomy and surgery. He also studied pharmacology closely; there was an esteemed tradition of

---

* Many of the ruins of this city, including the Altar of Zeus, are now in the Pergamon Museum in Berlin.

producing drugs in Egypt and he was able to get hold of accurate, uncorrupted versions of medicinal recipes. Galen wrote extensively about the local plants and food; he describes going down to the city harbour to talk to sailors about getting hold of drugs from even further afield. This proactive, practical approach to medicine would characterize his whole career.

Galen began writing as a teenager, and this explains, in part, at least, his extraordinary output—"fatiguingly diffuse," as one historian put it—some three million words that are collectively known as the Galenic corpus.[13] Astonishingly, this makes up around half the surviving literature of ancient Greece, but is only a fraction of the ten million words he is estimated to have written. Like Euclid and Ptolemy, his ability to survey and assess the theories he inherited from his medical predecessors, and to present them in a coherent and accessible form, is what makes him so important.[14] Unlike Euclid and Ptolemy, however, Galen did not do this in one large, convenient (particularly so, for our purposes) volume. His monumental contribution to the discipline of medicine is spread out in hundreds of separate books on a huge range of subjects. About a fifth of them are commentaries on Hippocrates (460–370 BC), the titan of ancient medicine, who provided the basis for Galen's own work. Hippocrates' "fourfold scheme," of the humours, elementary qualities (earth, air, fire and water), seasons and age, was the inspiration for Galen's own system. Humoral pathology, the idea that the human body contains four fluids—black bile, yellow bile, phlegm and blood—and that an imbalance in these humours causes disease, was the prevailing medical doctrine well into the nineteenth century. Galen also made important discoveries of his own. He was the first to prove that the arteries carry blood, transforming knowledge of the circulatory system. He explained the difference between types of nerves and used pioneering surgical techniques. His greatest pas-

5. Anatomical votives found at the Temple of Asclepius in Athens, carved to represent different body parts. These would have been offered to the god Asclepius in the hope he would cure patients of a particular malady.

sion, though, was anatomy, and although he was not able to dissect human cadavers (the practice had been illegal in the Roman Empire since 150 BC), he transferred the knowledge he gained from dissecting pigs and monkeys, something he often did publicly, in front of a rapturous audience. The resulting theories were not challenged until Andreas Vesalius published his revolutionary work on anatomy in 1543.

At the age of twenty-eight, Galen returned to his native Pergamon to become a doctor at the gladiatorial school there. By then, he had spent ten years studying—"the longest medical education on record"[15]—giving him a unique overview of the subject. Gladiators were the elite athletes of the day, and, while treating their wounds, Galen gained insights into the workings of the nerves and muscles, building up practical experience that would stand him in good stead later on in his career. He developed new ways of suturing deep tissue muscle and created innovative therapies for the treatment of injuries.

In AD 161, he moved to Rome, where he quickly gained a reputation as a talented healer. He continued writing books, he revised works he had written earlier, and began to give public lectures and anatomical demonstrations. Galen revelled in his role as a respected member of the Roman elite, wealthy, cultivated and well connected, yet he remained something of an outsider, a Greek in the Roman world. He always wrote in his own language[*] and was sometimes scornful of imperial society, especially its attitude towards scientific studies, complaining that:

> The materialism of the rich and powerful . . . who honour . . .
> pleasure above virtue, consider of no account those who possess
> some finer knowledge and can impart it to others . . . But the
> respect they give to men of learning corresponds only to their
> practical need of them. They do not see the particular beauty
> of each study and they cannot stand intellectuals. Geometry
> and arithmetic they need only in calculating expenses and in

[*] This was not unusual. The Roman state was bilingual, with Greek and Latin spoken by the majority of the elite.

improving their mansions, astronomy and divination only in
forecasting whose money they are going to inherit.[16]

This damning assessment of Roman intellectual life goes some way
towards explaining why scientific ideas made so little impact on
Western Europe in late antiquity and beyond. Translations of Greek
scientific texts into Latin were rare, and almost always abridged and
condensed as part of encyclopaedias. Some of Galen's works were
translated into Latin (he did, after all, spend long periods in Rome),
but the ones that weren't fell out of use as the Greek language gradually disappeared after the empire split in the fifth century. In the east,
as monasteries became the primary centres of knowledge and book
production, Galen's philosophical texts—distasteful to Christian
minds—also fell into obscurity. His medical works, however, with
their direct, practical relevance, were the most likely to be recopied
and shared, and, as we will see in the next chapter, they were initially
preserved by Christian communities in Syria and Persia, before their
discovery by Arab scholars in the ninth century.

Galen's voracious intellectual appetite wasn't limited to medicine. His early education in Pergamon instilled in him a passion for
philosophy, and one of his most profound achievements was synthe-
sizing Aristotelian ideas with medical thought. He developed this
theme in various works, most obviously *That the Best Physician is
also a Philosopher*, but also *On the Function of the Parts*, his massive
treatise on anatomy, both of which decisively unite philosophy with
medicine. Galen was also fascinated by philology, the historical study
of language, and lexicography—necessary areas of expertise for a
man who collected a huge library of manuscripts that he translated
and edited himself.

The sheer volume of Galen's writings worked against him when
it came to transmission. They were simply too numerous for them
all to survive. He realized this and responded typically by produc-

ing yet more works, detailing which of his treatises were the most important and how they should be read. The medical school of Alexandria solved this problem by organizing twenty-four of his treatises into the "Galenic syllabus"—confusingly, later called the "Sixteen Books." Students had to read them in a particular order, and the result was a concise yet comprehensive medical education. The Galenic syllabus was so successful that it spread to Syria and Italy, and would form the basis of medical study throughout the Muslim world from the tenth century onwards. According to Vivian Nutton, pre-eminent expert on Galen and ancient medicine in general, "its importance cannot be over-estimated."[17] It defined the study of medicine for centuries and demanded that doctors understood the principles behind the science, simultaneously helping to professionalize the discipline and impose rigorous standards.

No one was more vocal about his own fame than Galen himself—he claimed to have requests from patients living all over the empire—and he dictated treatises to twenty scribes at a time; this was textual production on an industrial scale, so it is not surprising that the geographical distribution of his work was so extensive. In the years after his death in Rome, in c.210, Galenic treatises were being copied as far away as Morocco, and he dominated the encyclopaedias of late antiquity. The size of the Galenic corpus, as with *The Elements* and *The Almagest*, forced scholars to come up with ways of processing and abridging the information. Producing manuscripts was time-consuming and expensive, so new copies of these works in their entirety were rare. Scientific ideas were condensed and transmitted via secondary literature—encyclopaedias, commentaries, dictionaries and summaries—not in the form their writers had intended, but at least some of the basic ideas survived and were passed down.

By the end of the fifth century, the map of knowledge had changed dramatically. Most of the ancient centres of learning had declined, schools had closed, libraries had been ransacked and burned, or left to decay quietly. Alexandria was still a centre of trade and of ideas, but the Library was a shadow of its former self. In 415, a mob of Christian zealots had murdered the philosopher and mathematician Hypatia. Believing her to be a witch, they flayed her alive with oyster shells. Then they turned their attentions to the magnificent Temple of Serapis and its collection of scrolls, taking "apart the temple's very stones, toppling the immense marble columns, causing the walls themselves to collapse."[18] This was a dramatic victory for Christians in Alexandria. The Serapeum had been the centre of pagan learning and power, its destruction was emblematic of the widespread "war waged by Christianity against the old culture and its sanctuaries: which meant, against the libraries."[19]

The tables would be turned yet again, two centuries later, with the arrival of the Arabs, in 641. By then, there was not much left of the Library, and its collection mostly consisted of writings on Christian themes, which were of no interest to the Muslim conquerors. Legend has it that the Caliph ordered all scrolls except those by Aristotle to be sent to the bath houses, where they fed the stoves that heated the bath water. Apparently, it took six months to burn them all. This is a good story, but the truth is both less dramatic and less entertaining. The most likely fate of the Library was gradual deterioration, the ink fading as the papyrus crumbled into dust. If no one thought to make new copies, loss was inevitable. The Great Library was gone, but its monumental reputation would become an everlasting symbol of both the power of knowledge and the tragedy of its loss.

By AD 500, Alexandria was on the wane, overtaken in size and political importance by Constantinople, the capital of the Eastern Roman Empire. Sixth-century emperors were more preoccupied with adorning their capital with grand buildings than tending to the imperial library, which had been founded in the mid fourth century.

None of the empire's rulers in this period were interested in scientific learning in the way the caliphs, who we will meet later, would be, or indeed in the way the early Ptolemaic kings had been. There must still have been small private collections in Constantinople, and there were certainly copies of Euclid, Ptolemy and Galen in the imperial collection a couple of centuries later. But, in the year 500, the fate of the three great strands of ancient science was uncertain. Only a few copies of *The Almagest* and *The Elements* remained, along with texts chosen from Galen's huge canon, scattered across Egypt, Syria, Anatolia and Greece. Some languished, forgotten, in the ruins of ancient temples or hidden in old chests in neglected libraries. Others could be found in certain monasteries or shelved in private collections, protected by a tiny handful of scholars who managed to keep the flames of astronomy, maths and medicine burning until the next great period of scientific endeavour dawned, in Abbasid Baghdad.

THREE

Baghdad

Baghdad, in the heart of Islam, is the city of well-being; in it are the talents of which men speak, and elegance and courtesy. Its winds are balmy and its sciences penetrating. In it are to be found the best of everything and all that is beautiful. From it comes everything worthy of consideration, and every elegance is drawn towards it. All hearts belong to it, all wars are against it and every hand is raised to defend it. It is too renowned to need description, more glorious than we could possibly portray it, and is indeed beyond praise.

—Mukaddasi,
The Best Divisions in the Knowledge of the Regions

IT IS SPRINGTIME, AD 917.

A group of ambassadors arrive in Baghdad, sent by the Byzantine Empress Zoë, from her capital, Constantinople. They are here to negotiate the terms of a peace treaty—the Byzantine and Muslim empires having been fighting for centuries over their shared border, running east to west across the Anatolian peninsula. The ambassadors are installed in one of the city's many palaces, where they spend two months waiting for their hosts to prepare their reception. The Muslim ruler of Baghdad, al-Muqtadir (895–932), the eighteenth caliph of the Abbasid dynasty, orders the redecoration of the entire palace complex in their honour. Furniture is rearranged, hundreds of curtains are hung, beautiful woven rugs from across the empire are laid, saddles polished and gardens pruned.

The day of the reception finally arrives . . .

The first palace the ambassadors would have entered was the Khan al-Khayl (Khan of the Cavalry), with its magnificent marble columns. "On the right side of the court were five hundred horses girded with gold and silver saddles, but without saddlecloths, and to the left were five hundred horses with brocade saddlecloths and long blinders. Each horse was entrusted to a groom attired in beautiful garb." From here, it was on to the zoological garden, with its "herds of wild animals . . . which drew near to the people, sniffing them, and eating from their hands." In the next courtyard were "four elephants adorned with brocade and cloth marked by figure work. Mounted on each elephant were eight men from Sind [in India] and fire hurlers." Al-Khatib al-Baghdadi (1002–1071), who recorded this description in his history of Baghdad, notes gravely that "this sight filled the Ambassadors with awe"—which was, of course, the aim of the enterprise.[1]

Al-Khatib goes on to describe, in breathless detail, a courtyard containing a hundred lions, muzzled and held by their keepers, and gardens with a pond and a stream lined with polished lead, shining white like silver, four golden boats with brocade seats floating on it. The gardens around were full of exotic trees, including 400 palms, whose trunks were ringed with gilt copper. Hundreds of craftsmen must have been employed to create these wonders, showcasing the full glory of Arab metalwork and artistry for the Byzantines. Then came the most amazing sight of all: "The Tree Room where a tree stood in the centre of a large round pond of limpid water. The tree had eighteen boughs, each containing numerous twigs on which were perched gold and silver birds of many species. The boughs, most of which were made of silver, though some were of gold, swayed at given times, rustling their leaves of various colours, the way the wind rustles the leaves of trees; and each of the birds would whistle and sing." This must have been a truly magical vision—a combination of incredible craftsmanship and ingenious mechanics that displayed the achievements of Baghdadi culture to the full.

The next palace was less subtle in its message; its walls were hung with thousands of pieces of armour—golden breastplates, leather shields, ornamental quivers and bows—and its corridors were lined with a countless number of slaves of different races, handpicked to demonstrate the breadth of Muslim dominions. After an exhausting tour of no less than twenty-three palaces in the sweltering July heat, alleviated only by occasional drinks of sherbet and iced water, the ambassadors were finally led into the presence of the Caliph al-Muqtadir.

They found him sitting on an ebony throne upholstered in cloth of gold, flanked by five of his sons.

This journey through the corridors of Islamic power was designed to show the Byzantine ambassadors that the Abbasid caliphate was still a force to be reckoned with, even though it had lost large chunks of its former territories—at its height, the Islamic Empire had stretched from the Atlantic coast of Africa to the Himalayas. It could still conjure lions, elephants and fire breathers from India; it could still put on a show. And its capital city, Baghdad, was still an important centre of scholarship.

But it was in decline. Only a century earlier, Baghdad had been at the peak of its golden age, unrivalled anywhere in the world for its beauty and sophistication, scholarship and wonder. In the heady first century of Abbasid rule, the caliph, as God's deputy on earth, was an exceptionally splendid and powerful ruler. In this early period, three caliphs made a particular impact: Al-Mansur (714–775), the second ruler of the dynasty, who founded Baghdad and became an inspirational patron of scholarship; his grandson, Harun al-Rashid (763–809), most famous now for the entertaining yet largely fictional depiction of his adventures in the *One Thousand and One Nights*, who was a fearsome warrior and a global leader, but also a passionate supporter of scholarship; and finally, Harun's son, the Caliph al-Ma'mun (786–833), under whose aegis Baghdad attracted the greatest minds of the day, and who, through a combination of wealth, enlightenment, curiosity and ambition, propelled human knowledge forwards.

If they had stepped out of the rarefied atmosphere of the palace, the Byzantine ambassadors would have found themselves in a city teeming with life, as around half a million people—Arabs, Persians, Turks, Bedouins, Africans, Greeks, Jews, Indians and Slavs—jostled for survival and success. They ate, slept, prayed and worked side by side in the biggest melting pot on earth. Many had been brought to the city as slaves—human cargoes in a trade that was even more lucrative than silk.* Those who came of their own free will were often pursuing a dream: merchants hoping to make their fortunes; singers hoping to make their names; scholars hoping to make discoveries. This exceptional mix of races, languages and faiths had a vital effect on the scholarship of the city—books were translated so that the knowledge they contained was opened up to other cultures, while scholars were able to bring their own ideas and traditions to the work of others. Cultural exchange happened freely in Baghdad, resulting in an explosion of knowledge of all kinds—in the sciences which are the main focus of our story, but also in theology, political theory, philosophy, law, history, literature and, above all, poetry. For the caliphs and their courtiers revered poetry, which was often sung, as part of the dynamic oral tradition that characterized Arabic literature. The slave girls and singers who performed and played were rewarded with great fame, adoration and wealth.

The Abbasid caliphs were united in a devotion to their God, and their role as "Commander of the Faithful." Intertwined with this was the belief that their dynasty was predestined to rule, and to do so for centuries. It is difficult to appreciate how fundamental preordination was to medieval men and women across both the Muslim and the Christian worlds. In a world of untold danger and confusion, it was important to feel that whatever you were doing was part of a divine plan—this was as true for farmers planting their crops or merchants setting out on a voyage as it was for caliphs commanding an empire. There were various ways of discovering whether whichever god you worshipped approved of your plans, but the night sky was the most

---

* As Peter Frankopan discusses in his book *The Silk Roads*, the demand for slaves in this period was enormous, with huge numbers of people being captured, transported and then sold into servitude.

powerful. From the beginning of civilization, mankind had been captivated by the stars, and the people of medieval Baghdad were no exception. During the long, sultry nights, they lay on the flat roofs of their houses, gazing up at the glittering firmament, imbuing the stars with monumental significance. As the Caliph al-Ma'mun himself explained:

> Sleepless I watch the heavens turn
> Propelled by the motion of the spheres;
> Those stars spell out (I know not how)
> The weal and woe of future years.
> If I flew up to the starry vault
> And joined the heavens' westwards flow
> I would learn, as I traversed the sky,
> The fate of all things here below.[2]

In a terrifyingly uncertain world, the stars offered a map of the future, both a glimpse into the mystical world of heaven and the key to discovering the secrets of the earth below. Today, we make a distinction between astronomy, the study of the stars and planets, and astrology, the interpretation of their influence on human affairs. In medieval Baghdad, no such distinction was made between the two disciplines. People believed that astrology could predict the weather, natural disasters and plagues, but also their own personal health, fortune and characteristics (through horoscopes)—they used it to make decisions about every area of their lives. Astrology was a bridge between the human world and the divine, the known and the unknown.

As we peer back through the centuries to early medieval Baghdad, it is difficult to gain a very clear picture of exactly what went on there, but the prolific output of contemporary Baghdadi historians helps, none more so than that of the great Persian scholar al-Tabari, whose *History of the Prophets and Kings* describes the deeds of the caliphs in often stultifying detail. Running to an exhaustive thirty-eight volumes, it is one of the most balanced sources of information for the period up to 915, if not the most entertaining read. Much more interesting is *The Fihrist*, by Ibn al-Nadim, "a catalogue of the books of all peoples, Arab and foreign, existing in the language of the Arabs, as well as of their scripts, dealing with various sciences, with accounts of those who composed them."[3] Written at the end of the tenth century, it is a major source of information on all kinds of knowledge, writers and scholars in the Arab world at the time. As the son of a Baghdadi bookseller, Ibn al-Nadim grew up surrounded by books and scholars, and, in *The Fihrist*, he paints a vivid picture of scholarship in the golden age. His extraordinary book is the result of a lifetime spent gossiping and researching the intellectual milieu of his city.

By the time scholars were eagerly perusing *The Fihrist* in the bookshops of Baghdad, almost the entire corpus of ancient knowledge—Greek, Egyptian, Indian and Persian—had been recovered, translated into Arabic and critically edited. At a time when many Europeans were living on turnips and trying to fend off the Vikings,* scientists in Baghdad had measured the circumference of the earth, revolutionized the study of the stars, developed rigorous standards for translation and methods for scientific practice, produced a map of the world, advanced the basis of our modern number system and defined algebra, founded new disciplines in medicine and identified the symptoms of several diseases. In a surprisingly short time, the Abbasids and their subjects had redrawn the map of knowledge and made Baghdad into an important centre

---

* There was, of course, cultural and scribal activity at the court of Charlemagne and at certain monasteries in this period, but no scientific study of any significance.

of scientific study, bathed in the glow of a golden age of discovery and enlightenment.

Just two centuries before, Baghdad had been a tiny Persian village clustered around a Nestorian monastery on the banks of the River Tigris.[4] Marked out for greatness by its geographical position, just where the Tigris and the Euphrates flow close to one another, Baghdad lay in the heart of the ancient, fertile land of Mesopotamia, variously described by contemporary writers as the "navel" or "crossroads of the universe." In the course of five or so millennia, people had settled here, planted their crops and dug irrigation channels. Agriculture flourished, and food was so plentiful—at times, the land produced three crops a year—that many believed it was the location of the Garden of Eden. The huge wealth generated by this profusion had formed the basis for a series of empires in the lush green basin between the rivers: the Babylonians, Sumerians, Assyrians and Achaemenids, the Macedonians under Alexander the Great and the Seleucid dynasty. In 150 BC, it became part of the Parthian Empire, in whose hands it remained until AD 224, when a Persian dynasty called the Sassanians arrived. They would rule over the area until the mid seventh century.

Bedouin tribes from Arabia had been descending sporadically on the settled peoples of Mesopotamia for centuries, abandoning their harsh nomadic lifestyle in search of something better, but in the mid seventh century something happened that would transform the course of history. In a cave, high above the city of Mecca, the Archangel Gabriel appeared to a forty-year-old merchant called Muhammad, imparting revelations to him that were later written down as the Qur'an. The religion of Islam was born. Over the next few decades, Muhammad's following grew exponentially as he spread his message first in the city of Medina, then Mecca and the rest of the Arabian peninsula, bringing its disparate tribes under a single banner and providing them with a single unifying creed by which to live. Muhammad died in 632, but his followers, fuelled by the fervour of their new faith and carried at great speed by their magnificent horses, swept into Egypt, Syria and Persia. Their timing was perfect. The once-mighty Sassanian Empire was on the brink of collapse, weakened by decades of war with its Byzantine neighbour.* In many cases, all the Muslims had to do was arrive at the gates of a city and look menacing in order for the citizens to surrender and offer vast sums of money as tribute. Egypt fell quickly and, within a matter of decades, the Muslims had overrun the entire Middle East and beyond, defeating the Sassanians and taking huge tracts of land from the Byzantines. The revenue that flowed into their coffers from these conquests was staggering, wealth that fuelled the rampant extravagance for which the Muslim elite became famous. By the early eighth century, they had conquered an area larger than the Roman Empire at its height—some five million square miles. And, for the first time in a thousand years, the lands that had originally been united by Alexander the Great were again under one ruler.

The cultural significance of this was huge. Alexander, as Basileus of Macedon, Hegemon of the Hellenic League, Shahanshah of Persia, Pharaoh of Egypt, Lord of Asia and, perhaps most importantly, pupil of Aristotle, had spread the language, philosophy, religion and traditions of Greece across his empire, founding outposts of Hellenism as far away as the border with China. With characteristic modesty, he named just twenty of these cities after himself, creating a network of Alexandrias that hummed with Greek culture for centuries after his death. This "wide movement of cultural diffusion" was slow, but it was also extremely long-lived, lasting almost a

---

* At this point, the Sassanian (Persian) Empire spanned Iran, Iraq, Syria and the Caucasus, and stretched north into the vastness of the Central Asian steppe and eastwards to the mountainous border with China.

thousand years.[5] Almost a millennium after he died, many of Alexander's cities—Merv, Aleppo, Alexandria, Bactria, Baalbek—shone brightly on the map of knowledge, and, because they were all in regions now ruled by the Abbasid caliphs, they fed the Baghdadi golden age with ideas, scholars and books.

In the decades following the Muslim conquests, thousands of Arabs migrated northwards, settling in Iraq, Iran and in the vast province of Khurasan—fertile, prosperous, and home to the splendid cities of the Silk Road, such as Balkh, Merv, Nishapur and Samarkand. However, the rulers of the new Muslim Empire soon discovered, like Alexander before them, that it was impossible to rule by force—there were too few of them, their numbers stretched too thinly across the massive territories they had conquered. They tolerated their non-Muslim subjects and taxed them in accordance with Islamic law. They encouraged continuity so that local populations stayed and farmed, and they adopted existing systems of government, employing, emulating and befriending many members of the Sassanian elite.

It helped that Sassanian culture was one of the most sophisticated and impressive on earth, and that Arab culture was young and relatively primitive. Just a few generations earlier, Muhammad's people had been Bedouins, wandering the deserts of Arabia. Now, they were rich beyond their wildest dreams, and they wanted the lifestyle to go with it. The Sassanians ate exquisite food, lived in sumptuous houses and surrounded themselves with brilliant scholars, musicians and poets. The Arabs were entranced. They enthusiastically absorbed all the splendour that Sassanian Persia had to offer, fusing it with their own courtly traditions to create one of the most extravagant, extraordinary cultures the world has ever seen. Baghdad became the ultimate expression of that fusion, thanks to the visionary caliphs of the Abbasid family.

In the years leading up to 750, the Umayyad dynasty ruled the

Muslim Empire, but the hairline cracks that had begun to split Islam as soon as Muhammad died were deepening—the Sunni/Shia schism that has divided the Middle East ever since. As the Prophet's descendants squabbled over who had the best claim to power, the Abbasids (who traced their line back to Muhammad's uncle, Abbas) worked quietly to foment discontent. In 747, their leader, who went by the pithy epithet of al-Saffah ("Shedder of Blood"), unfurled his black flags over the city of Merv and unleashed a revolution. Having seized power, he went on to brutally hunt down and exterminate every member of the Umayyad clan in an orgy of slaughter that was vicious even by contemporary standards, allegedly culminating in exhuming bodies and burning graves. But one young Umayyad prince, Rahman, escaped and fled to Spain, where he founded a rival dynasty. His descendants went on to build a vibrant centre of learning in the city of Córdoba, as we will see in the next chapter. Al-Saffah's reign was as short as it was violent. When he succumbed to smallpox in 754, the caliphate passed to his brother, Abu Ja'far Abdullah ibn Muhammad al-Mansur (c.713–775).

Luckily, al-Mansur was a very different man from his brother. Tall, with a wispy beard and penetrating eyes, he spent his reign consolidating power and establishing stability. His greatest achievement was the foundation of a new capital city, which he called Madinat al-Salam—the City of Peace—the city we call Baghdad. In refocusing imperial power away from the Umayyads' Arab stronghold in Damascus, al-Mansur deliberately strengthened his connection with the Sassanian elite, rooting his new city in the glory of the ancient Persian heritage. This was both official and personal—al-Mansur's closest friend was a Persian from Khurasan, called Khalid ibn Barmak, whose family had provided vital support to the Abbasid revolution. Exotic and cultured, the Barmakids were from Balkh, in the far north of the empire, where they studied Aristotle and learned

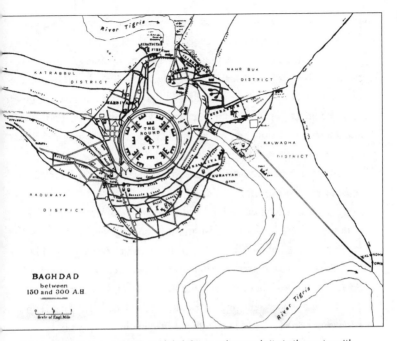

6. A reconstructed map of early Baghdad showing the round city in the centre with its four gates, and the various canals intersecting the districts around it. On the left, on the other side of the River Tigris, is the Shammasiya district, where al-Ma'mun built his observatory.

to read Greek; more than any other Persian family, they embodied the spirit of sophistication and brilliance that so dazzled their Arab overlords. Barmak helped al-Mansur find a site for his new capital. Together, they travelled southwards and chose the tiny village of Baghdad, just thirty kilometres north of the old Sassanian capital at Ctesiphon.

As al-Mansur explained to his generals, "Here is the Tigris, with nothing between us and China, for all that is from the sea can come to us on the river, as can the provisions of the Jazira, Armenia and the surrounding areas. And there is the Euphrates, on which everything from Syria, Raqqa and the surrounding areas may come."[6] Baghdad was ideally situated, with direct access, via the al-Sarat and Nahr'Isa canals, to the major trade routes: north-west, up the Euphrates to Syria and beyond; north-east, on the Tigris, via Mosul; and south, to the Persian Gulf, gateway to India, China and the Far East. Baghdad also lay at the centre of a vast network of land routes. Travellers and merchants from the East would have descended along the Silk Roads with their winding caravans, down through the mountains of Iran, on their way to North Africa, Arabia, Syria, the Mediterranean coast and on to Europe.

The stars also appeared to be aligned when al-Mansur and Barmak discovered that local Christian tradition had even predicted that one day a king called Miqlas would found a great city there. Happily, al-Mansur remembered that Miqlas had been one of his childhood nicknames, shouting in delight, "By God, I am that man!"[7] The Persian Jewish scholar Mash'allah, an Arab called al-Fazari and a Persian Zoroastrian called Nawbakht—leading lights of al-Mansur's team of court astrologers, whose colourful combination of races and faiths prefigured the multicultural character of later Baghdadi

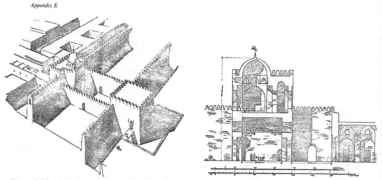

Appendix E

Figure 5. The Gates Reconstructed (Herzfeld, p. 127).   *Figure 6. Inner Gate Reconstructed (Herzfeld, p. 126).*

7. Modern reconstructions of the gates of Baghdad showing the double-wall structure and the two-storey towers.

Appendix E

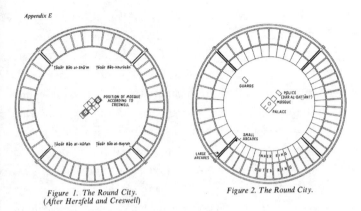

Figure 1. The Round City.
(After Herzfeld and Creswell)

Figure 2. The Round City.

8. Detailed map of the Round City showing the Great Mosque and the defensive double walls.

scholarship—drew up charts to determine the most propitious moment to start building the new city: 30 July 762, at two o'clock in the afternoon. Al-Mansur laid the first stone himself.

Al-Mansur sent out decrees to all corners of his empire, calling on the skills of thousands of workmen, surveyors, engineers, architects, blacksmiths, carpenters, builders and slaves to come and raise his vision from the dust. Using cinders to map out his extraordinary design on the ground, he created a unique circular city. Project-managing every last detail of its construction, he terrified his workmen by going through their expenditure with a fine-tooth comb, throwing them in prison if they failed to account for every last coin. Labourers were paid two or three grains of silver a day, master-builders 1/24th of a dirham. Each brick was weighed, every penny was counted—he might have been building the greatest city on earth, but he was not going to waste a single handful of clay. Tabari tells us that, years later, a brick was found from the original walls with its exact weight carved onto the side of it. Two massive concentric circular walls of kiln-fired mud bricks formed the city's four-mile circumference, with a moat around the outer one. A third wall, inside, framed government offices and houses. Four enormous double-domed gates led out to the corners of the empire: north-east to Khurasan and Rayy, north-west to Syria, south-west to Mecca and south-east to Basra and the Gulf. On hot summer afternoons, the Caliph himself liked to sit in the upper room of the north-eastern gate to enjoy the breeze and gaze out towards the distant province of Khurasan, whose people had helped bring his dynasty to power.

In stark contrast to so many of his descendants, al-Mansur was not a big spender. "He did not avoid the most extravagant generosity when there was something to be had in exchange, but he would refuse the smallest favour if granting it entailed loss." His parsimony resulted in a treasury of 14 million dinars and 600 million dirhams on his death.[8] To put this in context, there were around twenty silver dirhams to one gold dinar and Tabari reckoned that a sheep would cost about one dirham. Earning the nickname "Abul Dawanik," or the "Father of Pennies," al-Mansur shared the Abbasid traits of drive, vision and intensity, but he lacked the profligacy and hedonism for which his descendants Harun al-Rashid and al-Ma'mun were to become famous. A pious, miserly teetotaller who disliked music and hated parties, he would have been horrified by his progeny's sybaritic lifestyle, whether the racy tales of the *Arabian Nights* or the claim that Caliph al-Mutawakkil shared his bed with 4,000 concubines (though, presumably, not all at the same time). Instead, under the guidance of his friend Barmak, al-Mansur embraced scholarship. In 766, he built the great al-Mansur Mosque next to the Golden Gate Palace in the Round City, and it quickly became a magnet for scholars.

Just as they are today, mosques and their schools were centres of worship, but also led the way in education and scholarship, becoming places where local communities went to learn and to share and discuss ideas; they were places where books were stored and libraries developed. The Muslim love of learning and teaching stemmed straight from the teachings of the Prophet himself: "There is nothing greater in the eyes of God than a man who has learned a science and who has taught it to people."[9] For the first few centuries, science and faith were in harmony—the quest for religious truth not only fuelling intellectual enquiry on a sweeping, philosophical level, but also requiring answers to specific, practical demands—working out the exact direction of Mecca so that prayer mats could be positioned correctly, or knowing the times of day for prayers. Religious doctrine had not yet begun to calcify and build up the walls of conservatism that would stifle science in later centuries. For now, the Muslim reverence for The Book extended to all books; they were "the inexhaustible fountain of inner life."[10]

In the late eighth century, a new product arrived in Baghdad that would transform the world of books forever: paper. In 751, the Arabs had defeated the Chinese at the Battle of Talas, in modern-day Kyrgyzstan, deep in the vast Central Asian steppe. Two of the prisoners of war who had been taken back to Samarkand passed on the secret of how to make paper from hemp and other fibrous plants. The first paper mill in the Arab world was built in Samarkand, from where the idea gradually travelled down the Silk Roads, reaching Baghdad in 793.

At much the same time as paper arrived in Baghdad, there were also great technological advances in the manufacture of ink and glue, and in the techniques of bookbinding. Together, these developments ensured that books became both more beautiful and more durable. Calligraphy, illumination and miniature painting also flourished as more and more people were employed to meet the rising demands of the trade. The greatest of these were the warraqeen, or "traders in paper," who ran the bookshops. In the late ninth century, the scholar al-Ya'qubi counted over a hundred of them in the Waddah suburb of Baghdad alone. They had their own souk there, and many employed teams of scribes to produce the books they sold, which were displayed on trestle tables; browsing was encouraged. Many warraqeen were scholars in their own right and their shops became places where intellectuals gathered—unofficial academies and hotbeds of scientific debate. Some joined in the search for manuscripts, travelling far and wide to unearth the treasures of previous civilizations. The warraqeen also helped to make it possible for scholars to earn a living from writing, and, by developing the trade in books, they transported knowledge from Baghdad across the Dar al-Islam, as the Muslim sphere of influence was called. Without them, the enormous literary output of the Arabic-speaking world—there were over 5,000 Muslim authors writing by the end of the eleventh century—would not have been possible.

Within forty years of its foundation, Baghdad was a thriving metropolis. People came there from across the Dar al-Islam and beyond, attracted by the city's promise of tolerance and peace. The population rocketed and the city grew exponentially, creating practical problems like maintaining adequate sanitation, food supply and taxation. The empire needed infrastructure—roads, bridges, irrigation systems and canals—all of which relied on advances in technology and design. Even the most basic engineering projects required mathematical calculations. Medical knowledge was needed to help cure disease and save lives. Astrology was an integral part of everyday life, especially in medicine, where it was used in diagnosis. Astronomy was fundamental to astrology and to any kind of geographical investigation, navigation and the production of maps (which had their own obvious military relevance). None of these could be pursued without mathematics—the language of measurement, calculation and accuracy. Academic study and practical knowledge were entwined, fuelling the engines of cultural production and scientific endeavour.

As Arabic was formalized from an amorphous collection of oral traditions into an official, written language in the eighth century, a huge programme of translation began from Persian or Pahlavi (the written form of Middle Persian). Many of the first wave of books to be translated into Arabic were practical treatises on government, administration and taxation, but before long attention turned to the extensive Persian canon of astrology and astronomy. The stars played a huge role in Zoroastrianism, the state religion of Sassanian Persia; over the centuries, scholars had built up a sophisticated body of work on the subject, which incorporated ideas from India, Greece, Egypt and strands that stretched all the way back to the Babylonian civilization of 1800 BC.

Books were transported in and out of Baghdad and across the Arab world relatively easily. This process was helped by the governmental postal service, which operated throughout the empire, with messages carried by relays of camels, mules, horses and pigeons. The walls of the main office in Baghdad were hung with huge maps, used by travellers and pilgrims to plan their journeys. A network of caravanserais (inns), hospices and roadside water cisterns served the pilgrims, merchants, pedlars, soldiers, messengers, preachers and other travellers. During the long evenings, people would gather round caravanserai campfires to eat, drink and rest, but most of all to talk and exchange gossip—making them vibrant hubs of information, the newspapers and networking sites of their day. Travellers banded together, joining the long caravans of camels that transported goods across the empire—the safest way to confront the dangers of crossing the deserts of Arabia, North Africa and Iran. Among them were scholars prepared to journey hundreds of miles, risking sandstorms, disease, floods, brigands and wild animals in their search for books, propelled by the fear that ideas could be lost for ever. They travelled east across Persia and north to Anatolia in the Byzantine Empire, where Greek was still the main language. Here, too, were ancient cities with old temples and monasteries full of books.

In the 770s and early 780s, al-Mansur and Khalid ibn Barmak created the perfect conditions for science to flourish in Baghdad. Khalid's erudite son, Yahya, was appointed as tutor to Harun al-Rashid (Mansur's grandson) and—for the time being, at least—the destinies of the two families were bound together by love and mutual respect. When Harun became caliph in 786, he summoned his beloved tutor, telling him, "My dear little father, it was you who placed me on this throne, it was by your aid and the blessing of heaven—yes, by your happy influence and wise advice! And now I invest you with absolute power."[11] Yahya commissioned the first translation of *The Elements*, and soon the entire Baghdadi elite was following his enlightened example, pouring money into the recovery of ancient texts. Book production soared as texts were read out to roomfuls of scribes so that many copies could be made at once. Within a generation, no self-respecting palace in Baghdad was complete without a library bursting with books, staffed by scholars and scribes.

Harun al-Rashid was an intensely paradoxical figure: hedonistic, energetic, violent, pious, generous, cruel and clever. The extravagance of his court was legendary. His wife, Zubaydah, could give any modern billionaire a run for their money; she dined off gold and silver tableware, and started a fashion for adorning shoes with rubies. Her more flamboyant costumes were encrusted with so many jewels that, when she stood up, she had to be supported by a servant on each side. Her husband also lived each moment as if it were his last. Harun was as famous for making love as he was for making war, but his appetite for learning was equally impressive. Once in power, he put the mighty weight of the caliphate behind the search for ancient books. Like his ancestors before him, he would send raiding parties into the southern fringes of the Byzantine Empire as many as three times a year.[12] During the chaos of these skirmishes, soldiers grabbed anything they could get their hands on, and books were high on the list of valuable booty.[13]

In Baghdad, Harun founded the Bayt al-Hikmah—House of Wisdom—to house these books and the scholars who worked on them. We know tantalizingly little about what it looked like, how it functioned or where it was, so any suggestions have to be based on descriptions of similar places, combined with a dose of conjecture and imagination.* However, we do know that it contained a library—a room or rooms that housed the many books that were collected and produced there—so there must also have been places where scribes copied out the manuscripts, and where scholars worked on translations. There were probably a considerable number of staff employed to run it—*The Fihrist* mentions some librarians and directors by name and lists many scholars who worked there— but the scores of messengers, fetchers and carriers, cleaners and so on who supported them can only be assumed. Other centres of knowledge of the same period usually offered scholars somewhere to sleep and provided food, so there would have been rooms for eating and socializing, traditionally furnished with rugs and low tables. They often stored the books in chests rather than on bookshelves and had desks at which to consult them, with paper and reed pens provided for free. It is impossible to know whether the House of Wisdom was situated inside one of the vast palace complexes or if it was separate, but, at the very end of the ninth century, Harun's descendant, Caliph al-Mutadid, began building a new palace which had an annexe with accommodation and study rooms for scholars. He also seems to have planned to move al-Ma'mun's library there. Was this annexe a copy of the House of Wisdom? Or was it an improved design, positioned closer to the heart of the palace? Scholars probably led fairly fluid lives, working in the House of Wisdom long into the evening and

---

* There is no evidence for where the House of Wisdom actually was, causing some modern scholars to suggest that it only existed symbolically, in several places, rather than in one location.

sometimes spending the night there—if they were involved in a big project, say—then, at other times, going home to their wives and families elsewhere in the city.

In the village of Karkar, just outside Baghdad, a nobleman called al-Munajjim had in his castle an important collection of books, and his library was probably run in a similar way to the House of Wisdom. Scholars came from many different countries to study at al-Munajjim's library; he entertained them as his guests and, in return, they enhanced his reputation as an enlightened, erudite patron of learning. This was mirrored throughout the empire: the elite took the patronage of scholarship very seriously, looking after scholars and giving them everything they needed—writing materials, bed and board, money, books and academic encouragement—to make the best of their talents. All these elements must have been available in the House of Wisdom, as were the powerful forces of cooperation and competition—scholars working together to share their ideas and capabilities, but also striving to outdo one another—thus pushing the boundaries of knowledge ever outwards.

There were several libraries for public use in Baghdad, many of them attached to mosques and their madrasas;* they were essential to the Islamic cult of books and, thanks to generous bequests, they blossomed and grew.[14] They also made scholarship more accessible to the masses. For, even after the introduction of paper, books were still expensive—a large multi-volume text like al-Tabari's *History* could cost a hundred dinars. But, as far as the Arab elite were concerned, there was no better way of spending money. Baghdad was full of private libraries, which became popular meeting places for scholars and patrons, and unofficial centres of academic debate. A library became the ultimate status symbol.

---

* Madrasas are educational institutions that were (and still are) often attached to mosques.

Scholarship flourished in Baghdad under Harun's influence, but trouble was brewing and the first storm broke in 803. Harun—for reasons that aren't clear—suddenly turned on the Barmakids, throwing his former tutor, Yahya, now an old man, into jail and brutally murdering his son, Jafar. The mighty Persian family, who had danced on the knife-edge of royal approval so successfully, had fallen, never to rise again. Next, Harun addressed the problem of his succession, naming Abd Allah al-Ma'mun, his son by a slave girl, second in line to the caliphate after his legitimate first-born son, al-Amin, whose mother was the flamboyant Zubaydah. During the years leading up to his death in 809, Harun tried to ensure that his crown passed peacefully to al-Amin. But fate was against him. Al-Ma'mun, who went on to preside over the greatest era of Arabic intellectual endeavour of all time, was not a man who would put up with coming second. Following in the blood-soaked footsteps of his great-uncle al-Saffah (who had founded the dynasty), he established a power base in Merv, capital of the vast northerly province of Khurasan, his mother's birthplace, and plunged into a bloody civil war with his brother that culminated in the fourteen-month siege of Baghdad. Al-Amin didn't stand a chance. Al-Ma'mun was brilliant, charismatic and unstoppable.

Al-Ma'mun was crowned caliph in 813, but remained in Merv until 819, when he finally made the thousand-mile journey back to Baghdad. The city never fully recovered from the civil war and, although he was a strong leader, it was beset by outbreaks of violence and factionalism. Like his father, al-Ma'mun lived a life of unbelievable splendour and luxury. His palaces were furnished with the most beautiful objects and his courtiers were treated to the most fabulous banquets: reclining on silk cushions, they were serenaded by dancing girls as they feasted on figs, pistachios, grapes, pomegranates and saffron-infused baklava dripping with honey, all served by glamorous eunuchs. When not feasting and carousing, the court whirled on a merry-go-round of polo playing, hawking, fencing, hunting and horse racing.

This frenetic hedonism was matched by intense academic activity. The Barmakids had imbued al-Ma'mun with a fascination for the world and such a reverence for Greek learning that it was claimed Aristotle had visited him in a dream.[15] His intellectual curiosity was relentless. While on campaign in Egypt, he became obsessed with translating hieroglyphs and commissioned a local sage to transcribe them and try to work out their meaning. Of all the Abbasid rulers, al-Ma'mun had the greatest personal interest in science; soon after he returned to Baghdad, he re-established his father's House of Wisdom, making it into a centre of Arabic learning that heralded a new dawn of state-sponsored scientific progress. Al-Ma'mun set about collecting books on a grand scale, writing to "the Byzantine emperor asking his permission to obtain a selection of old scientific [manuscripts], stored and treasured in the Byzantine country," and then sending "a group of men . . . [including] Salman, the director of the Bayt al-Hikmah [House of Wisdom]," on an embassy to Constantinople to collect them.[16] By the mid ninth century, the library of the House of Wisdom was the largest repository of books in the world.

Knowledge flowed into Baghdad from every direction, and in various languages. The Christian Church in the Middle East was well established, its numbers swollen by Syriac Christians, who had split with the main Eastern Orthodox Church over doctrinal differences and founded their own Nestorian Church. Persecuted by the authorities in Byzantium in the fifth century, many had fled to the Persian Empire, where they had set up centres of Christian learning at Antioch, Edessa and, later, Nisbis—city of white roses, wine and scorpions. In these cities, Greek theology, philosophy, medicine and astronomy were taught and studied in Syriac—a dialect of Aramaic and the literary language of the Christian Middle East. Men like Theophilus of Edessa, chief astrologer at the Abbasid court, brought the works of Aristotle and other Greek philosophers with them to Baghdad. The Nestorian Christians had close links to ancient Greek scholarship and their expertise and early translations of Greek books from Syriac into Arabic were the foundation of scientific scholarship in Baghdad. The Syriac scholar Severus Sebokht (575–667) studied *The Almagest* and wrote a treatise on astronomy in which he recommended Ptolemy's work to anyone wanting to delve deeper into the subject.

As manuscripts were translated from Syriac and Pahlavi into Arabic, scholars in Baghdad began to realize the extent of ancient learning, and therefore how much lay out of reach. Al-Mansur himself had written to the Byzantine Emperor asking for scientific texts. It was no secret that many ancient Greek manuscripts lay hidden behind the fortified walls of Constantinople, a city that had eluded invasion and so preserved its ancient monuments and libraries. The Emperor responded by sending a chest of scientific books, including Euclid's *Elements*. In the following decades, scholars translated it into Arabic, initiating a rich tradition of mathematical study. The original copy has not survived, but there is a similar version, made about a hundred years later in Constantinople, which is now in the Bodleian Library. Its careful Greek script, with neat diagrams illustrating the mathematical hypotheses, has been annotated in the margins by its first owner, Arethas of Patrae, Bishop of Caesarea, as he tried to master Euclid's theorems. Al-Mansur's copy was the first, that we know of, to arrive in Baghdad. If there was an earlier version of *The Elements* in Syriac, it has not survived, and it appears that al-Mansur did not get his copy translated straight away; the first Arabic version was produced in the reign of Harun al-Rashid.

Mathematical ideas also came to Baghdad from the East. In 771, a traveller arrived in the city with a copy of a work of Hindu astronomy called the *Siddhanta* (*The Opening of the Universe*), by the Indian mathematician Brahmagupta (598–668). Unlike Euclid,

Brahmagupta did not set out his mathematical propositions clearly with proofs, but obscured them (as was traditional in Indian mathematics) under a veil of poetry—beautiful, but extremely difficult to unravel. Al-Mansur gave his court astrologer, al-Fazari, the Herculean task of translating the *Siddhanta*, which introduced Baghdad to the concept of "positional notation"—the way we write numbers to this day, using the digits 1 to 9, in columns of units, tens, hundreds and so on. The possibilities that this system opened up were limitless; when it was eventually adopted, it transformed the entire discipline of mathematics by allowing calculations that would have been impossible with the old Roman-numeral system. Positional notation was already known in Syria and had been admired by Severus Sebokht, who wrote about the "nine signs" of Indian mathematicians in 662.

In another groundbreaking part of the *Siddhanta*, Brahmagupta listed the rules of zero—mystical symbol of nothingness, the "fulcrum between negative and positive" that "unlocks the secrets of the universe."[17] This fundamental mathematical idea had developed gradually in several different incarnations and places. Ironically, the first written symbol for zero that we know of was carved onto a wax tablet, in about 3000 BC, in Sumer, just up the road from Baghdad. In this case, the zero is used as a simple symbol (two diagonal wedges), a place holder, to denote a gap in a series of numbers.

The idea gradually spread. Zero as a place holder became a common tool in bookkeeping, scrawled on receipts made out of pieces of bark, stuffed into merchants' saddlebags and carried between the bazaars of the Silk Roads and through the ports of the Persian Gulf. In India, it was gradually transformed from a useful accountancy symbol denoting the absence of something into the universal idea of nothing and a number in its own right. The concept zero itself originates in the Indian word *sunya*, meaning "void," a fundamental concept in Buddhist philosophy. In India, a fascination for huge numbers and the idea of infinity among the followers of Jainism propelled mathematics upwards into the realm of philosophy, where, free from the constraints of everyday necessity, it became abstract.[18] And numbers, no longer merely representing so many camels or apricots or grains of silver, became entities in their own right.

While Indian mathematicians grappled with abstract numbers of inconceivable magnitude, the ancient Greeks had been in love with geometry, working out mathematical problems by measuring lengths and drawing diagrams, rather than counting numbers. They had little use for the zero, or, indeed, for arithmetic, the tools of common traders. But the Muslim Empire was founded on commerce— Muhammad himself had been a merchant—and so Baghdadi scholars did not share this prejudice. In the ninth century, one of the greatest of these was Muhammad ibn Musa al-Khwarizmi (shortened by Latin writers to Algorithmus), a Persian genius who developed the concept of the algorithm in his *Kitab al-Jebr*—the work that, in 820, established algebra as an independent discipline for the very first time. This book was designed to help people solve everyday problems, working out their taxes or dividing up land for irrigation, for example. It did so by taking the mathematics to a higher level of abstraction and developing general rules that could be applied to many different questions, making it extremely useful to a wide range of people. Among other things, al-Khwarizmi listed, for the first time, several types of quadratic equation and provided methods of solving them. Like so many other scientific texts written in Baghdad, it was copied many times and taken out across the Islamic Empire.

We know almost nothing about al-Khwarizmi's life, but his name means "of Khwarezm," a province far away on the parched shores of the Aral Sea. At some point, his people had made the long journey westwards along the ancient Silk Road from Nishapur, where the most delicious pistachios and pomegranates grew, through the Zagros Mountains and down into the verdant orchards and vegetable gardens on the outskirts of Baghdad. Here, he lived and worked with his friend, the "philosopher of the Arabs," Abu Yusuf Ya'qub ibn Ishaq as-Sabbah al-Kindi (801–873).[19] Both men wrote books about the Hindu-Arabic decimal system, extolling its beautiful simplicity and boundless potential. But they weren't at all successful in getting people to abandon the old ways of counting— huge paradigm changes like this always take time—and the Babylonian sexagesimal system, which revolved around the number sixty (hence the minutes on our clocks), remained the norm, along with Greek and Roman numerals, for centuries to come.

The *Siddhanta* probably came to Baghdad via the city of Jundishapur, in modern-day Iran, which had been a centre of medical studies for several centuries, a place where scholars were able to meet and combine ideas from Greece and Egypt with traditions from the Far East. In the third century, the Persian scholar-king Shapur I had brought his new Roman wife to live in Jundishapur, accompanied by her two Greek physicians. They had taught the theories of Galen and Hippocrates, and had thus made the city a centre of med-

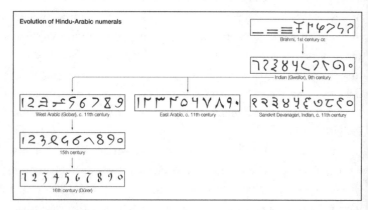

9. The development and geographical movement of the Hindu-Arabic numerals, AD 100–1600.

ical study and practice, with a hospital, academy and library. After 529, Greek philosophers arrived from Athens, fleeing persecution by the Byzantine Emperor, while Nestorian Christians also came and built a community there, bringing ancient Greek manuscripts with them as they migrated eastwards. In the sixth century, the Sassanian king, Khusraw, sent one of his doctors to India and China to invite scholars to come to Jundishapur and exchange medical ideas. These ideas were combined with those of the Jewish, Persian, Greek and Syriac traditions. This fusion of different strands of medical thought was brought to Baghdad by al-Mansur, who, suffering from a bad stomach ache, summoned the elegantly named Nestorian doctor Jurjis ibn Jibril ibn Bakhtishu (Persian for "Saved by Jesus") from Jundishapur. Bakhtishu cured the Caliph's stomach ache, stayed in Baghdad and founded a dynasty of royal doctors. He brought the full wealth of Jundishapurian medical theory to the city, making Baghdad its successor in the development of medicine, and went on to be an important patron of learning and translation in the city.

Bakhtishu's grandson, Jibril, was appointed court physician in 805, a position he held intermittently for three decades, serving various caliphs, including al-Ma'mun. The Bakhtishu family was instrumental in bringing classical Greek knowledge to the fore in Baghdad, and it was during the reign of al-Ma'mun that this process reached its apogee, allowing Arabic science to move beyond ancient learning and become a tradition in its own right. Al-Ma'mun's personal curiosity and vision was one of the great drivers for this process. It was with al-Ma'mun's encouragement that al-Khwarizmi wrote his treatise on algebra. A follower of Mu'tazili theology, al-Ma'mun adopted various policies to establish himself as "God's Caliph" and ensure total religious, as well as political, power. He established a *minha* (inquisition) to arbitrate and enforce Islamic ideology and pursued the Mu'tazili tradition of using Greek philosophical texts and methods of dialectic (reasoning), creating an environment of enquiry and study. "He had jurists and the learned among men of general culture attend his sessions; he had such men brought from various cities and stipends for them allocated. As a result, people developed an interest in conducting theoretical investigations and learned how to do research and use dialectic."[20]

In al-Ma'mun's House of Wisdom, all the many intellectual traditions that had arrived in Baghdad converged as the scholars he employed translated, assimilated and built upon them, redrawing the map of knowledge. Only a few were ethnically Arabs; many were Persians—some Christian, some Zoroastrian—and many converted to Islam as a way of assimilating with the elite and fast-tracking their careers. Academic endeavour flourished in this atmosphere of wealth, technology, patronage and religious tolerance.

Al-Ma'mun was a demanding but visionary patron, breathtakingly arrogant, but childlike in his enthusiasm, constantly questioning and expecting the impossible of his scholars. Fortunately, he was surrounded by people with the imagination and intelligence to come up with answers, and none more so than the Banu Musa brothers. A brilliant, eccentric trio, they were the sons of one of al-Ma'mun's astrologers in Merv. When their father died unexpectedly, al-Ma'mun took the boys under his wing, educated them according to the Greek curriculum and then brought them with him to Baghdad. Muhammad, Ahmad and al-Hasan put their excellent educations and considerable intellects to good use, applying their knowledge of mathematics to practical engineering projects—designing canals, bridges and irrigation systems. They became indispensable to the Caliph and happy to rise to the challenge of his most audacious demand: that they measure the world for him. This had, in fact, already been done. Ptolemy, using information from earlier astronomers, estimated the earth's circumference to be 180,000 stadia. But there were no clues as to how long a stadion was—a small, yet fundamentally important, detail. What the Banu Musa and

10. The Banu Musas' diagram of the self-trimming lamp they invented, from their *Book of Ingenious Devices*.

al-Ma'mun's astronomers did know, however, was that Ptolemy's calculations were based on the beautifully simple proposition that, if you could measure one degree on the ground of the spherical earth, then all you needed to do was multiply it by 360 to find the circumference. A team of the best astronomers was sent out to the flat plain of Sinjar in north-west Iraq. In the dead of night, they divided into two groups and walked in opposite directions—due north and due south—and, using the positions of the stars, stopped when they had measured a one-degree angle of the earth's curve. They then walked back towards each other, carefully measuring the distance they had travelled. Next, they took an average of the two amounts—56.6 Arabic miles (the equivalent of 68 modern-day miles)—which they multiplied by 360 to give a total for the circumference of the earth of 24,500 miles, just 400 miles off the 24,900 miles measured by modern science. This was an extraordinary achievement, especially given the crudeness of the instruments they were using. Of course, al-Ma'mun had no idea how close his astronomers had got to the actual total, and, determined to get as accurate an answer as possible, he sent out another group to repeat the experiment in the Syrian desert soon afterwards. The total they reached was higher and further from the actual measurement, but of course there was no way they could know for sure.

This exploit gives us a clear idea of the atmosphere in ninth-century Baghdad. The Banu Musa brothers and their peers gave free rein to their imaginations, putting the full power of their wealth and intellect into the pursuit of scientific discovery and excellence. The Banu Musa were particularly famous for their patronage of translation. They sent teams of agents out on manuscript-finding missions and spent a fortune on producing books. According to Ibn al-Nadim, they paid their translators 500 dinars a month (one dinar contained 4.25 grams of pure gold, so, based on today's prices, this works out at something in the region of £18,000),[21] "equivalent to the salaries of senior members of the bureaucracy and vastly more than those of an ordinary craftsman or soldier."[22]

The brothers also wrote their own works; the most celebrated was the *Book of Ingenious Devices*—a collection of one hundred mechanical inventions or adaptations, some frivolous and some utilitarian, including a windproof torch, a flute that plays itself, a spill-proof jar and a self-regulating oil lamp. All these devices used mechanisms that either harnessed natural energy, like gravity or flotation, or transferred force from one part of the machine to another—and all of them are still in use in some form or other. One of the most significant was the crankshaft, which the Banu Musa adapted from designs that were used in Roman times. This revolutionary technology reached Europe in the late fourteenth century and is a vital component in engines of all kinds today. The "Room of the Tree" so marvelled at by the Byzantine ambassadors was certainly based on technology originally designed by the Banu Musa. The *Book of Ingenious Devices* was widely read across the Arab world and their ideas would travel to Muslim Spain and, from there, translated into Latin, into Western Europe.

One of the most brilliant translators working for the Banu Musa was a young Nestorian Christian called Hunayn ibn Ishaq (809–873). Bilingual in Syriac and Arabic, he had left his home town of al-Hira,* south of Baghdad on the Euphrates, to study medicine with the haughty Jundishapurian doctor Yuhanna ibn Masawayh. Yuhanna eventually agreed to teach Hunayn, in spite of his disdainful view of him as an al-Hiran, because al-Hirans were traditionally bankers or merchants. However, Hunayn's relentlessly enquiring mind and endless questions drove Yuhanna mad, and he was thrown out. This setback did not discourage young Hunayn. He left Baghdad and "travelled through the land to collect ancient books, even going into the Byzantine country,"[23] learning Greek as he went. When he returned to Baghdad several years later, Hunayn could "recite Homer by heart" and had amassed an impressive collection of books and set to work translating them into Arabic, often via Syriac.[24] He quickly gained a reputation for excellence, accepting commissions from many Baghdadi patrons. Al-Ma'mun employed him both as a physician and as a translator. Presiding over a team that would eventually include his son and his nephews, Hunayn revolutionized the translation process, using his intimate knowledge of Syriac, Greek and Arabic to give the actual meaning of each sentence, instead of simply translating word for word. Translators needed high levels of specialist knowledge to achieve this; linguistic expertise alone was no longer enough. He instituted rigorous standards by working

---

* Al-Hira was an important Arab settlement in the Persian Empire in the pre-Islamic period. It was located south of al-Kufa, in south-central Iraq.

closely with other translators, so that each manuscript was checked and edited several times. Hunayn and his team developed an entire technical vocabulary to express complex scientific ideas in Arabic, revising and improving the texts they translated and setting a gold standard for translation.

Hunayn's other great innovation was to collect as many versions of a book as he could (often in different languages) and collate them all to produce the most authoritative edition. As he himself explained, in relation to Galen's *Treatise on Sects*, "I translated it for a physician from Jundisabur [Jundishapur] . . . from a very faulty Greek manuscript . . . Meanwhile a number of Greek manuscripts had accumulated in my possession. I collated these manuscripts and thereby produced a single correct copy. Next, I collated the Syriac text with it and corrected it. I am in the habit of doing this with everything I translate. A few years later I translated the Syriac text into Arabic for Abu Jafar Muhammad b. Musa [al-Khwarizmi]."[25] It is difficult to overstate the importance of Hunayn's translation methods and the texts he produced. They enabled scholars to gain a profound understanding of ancient ideas so that they could organize, assess, challenge and correct them, before using them as the basis for their own discoveries. Hunayn's translations became the standard versions of many Greek texts, which were passed on and then, in later centuries, translated into Latin. He translated 129 works by Galen, many of which he sought out and found himself. Of one of these, *De demonstratione*, he wrote, "I sought for it earnestly and travelled in search of it in the lands of Mesopotamia, Syria, Palestine and Egypt, until I reached Alexandria, but I was not able to find anything, except about half of it at Damascus."[26] Hunayn is best remembered as a translator, but he was also one of al-Ma'mun's court physicians and wrote several books of his own, including *Ten Treatises on the Eye*, "the first systematic textbook of ophthalmology,"[27] which featured one of the first anatomical drawings of the human eye. In his translation work, one of Hunayn's main aims was to create an effective syllabus of Galenic texts for teaching medical students in Baghdad, in particular his own son, Ishaq ibn Hunayn (830–910), who followed in his father's illustrious footsteps by becoming a prolific translator and physician at the Abbasid court.

Hunayn's legacy was further consolidated in the late ninth century, when Muhammad ibn Zakariyya al-Razi (854–925) arrived in Baghdad from Persia.* Using Hunayn's excellent translations of Galen and Hippocrates, al-Razi—or Rhazes, as he became known in Western Europe—established the medical disciplines of psychology and paediatrics. He helped to found hospitals in Baghdad and his native town of Rayy (now a suburb of Tehran) at a time when endowments for the care of the sick were becoming common in Islamic culture. He also developed proper rules for clinical experiments using control groups, emphasized the importance of medical training and made an influential attempt to classify chemical elements, creating a prototype periodic table. He also wrote prolifically on a huge range of subjects, from astronomy, geometry and alchemy to fruit, nutrition and spiritual medicine. He is most famous for two books: a treatise describing the differences between measles and smallpox that made accurate diagnosis, and therefore treatment, possible for the first time, and a vast encyclopaedia of medical information, called *al-Kitab al-Hawi fi al-Tibb* (or *The Comprehensive Book on Medicine*, known as the *Liber continens* in Latin). His devoted students collated it, after al-Razi's death, using his working files—twenty-three volumes' worth—and for centuries it was regarded as one of the most important medical books in existence,

---

* Iranians still celebrate Razi Day every year on 27 August and there are hospitals and institutes named after him across the country.

trusted throughout Europe, North Africa and the Middle East.[*] Al-Razi, like Galen before him, had gathered all the medical knowledge available and then assessed, organized and categorized it so that it was easy to use and apply. This "old man with a large sack-shaped head" was, according to the historian Ibn al-Nadim's sources, "generous, distinguished and upright," and "so kindly compassionate with the poor and the sick"—an inspirational doctor, who led by example, revolutionizing medicine with his ethical stance, practical ideas and scientific rigour.[28] Hunayn died in 873, but his nephews and his son, Ishaq, were prominent members of the community of scholars who continued to produce translations in Baghdad. Father and son had worked on Galenic treatises together, but Ishaq was ultimately more interested in mathematics, although he did write a history of medicine in relation to philosophy and religion—the first of its kind. His greatest contribution to the history of science was to produce new editions of *The Elements* and *The Almagest*, both of which were revised by his colleague Thabit ibn Qurra (d. 901), so are known as the Ishaq/Thabit versions. These supplemented but did not replace earlier translations of the two books by a scholar called al-Hajjaj ibn Yusuf ibn Matar (*fl.*786–830). While no details about al-Hajjaj's life and career have survived, his textual influence, in the form of his translations of both *The Elements* and *The Almagest*, has been profound. Al-Hajjaj made his, and probably the first, translation of *The Elements* during the reign of Mansur's grandson, Harun, before improving it in a new version a few decades later, when Al-Ma'mun was caliph. The Ishaq/Thabit translation of *The Elements* was superior to al-Hajjaj's because it was based on better-quality Greek manuscripts, which had presumably come to light since al-Hajjaj had died in 833. Both versions were widely dissemi-

[*] It was one of only nine books in the early medical library at the University of Paris.

nated throughout the Arab world, and soon scholars combined the two texts to create different variants. In fact, all the Arabic copies of *The Elements* we have today blend the two traditions—no pure versions survive at all. Among other things, Ishaq worked on versions of Archimedes' *On the Sphere and the Cylinder*, Menelaus' *Spherics* and Euclid's *Data* and *Optica*—these went on to form the *Middle Collection* or *Little Astronomy*, which were studied in between *The Elements* and *The Almagest*.

The first Arabic translation of *The Almagest* had been produced by al-Hajjaj early in the ninth century, complete with technical terms and corrections to many of the inaccuracies of the original. The al-Hajjaj edition of *The Almagest* and the Ishaq/Thabit edition form the two main groups of manuscripts that survive today, although many of them merge the two versions in various different ways. Clearly, therefore, the scribes who copied *The Almagest* in the centuries that followed often had more than one text in front of them as they worked, and when texts were being copied by hand, there were myriad possibilities for the resulting work. The idea of a stable, standard edition was unthinkable until the arrival of the printing press in the fifteenth century.

Ptolemy's system of the universe was brilliant in many ways and was not replaced for a millennium and a half, but it was full of inconsistencies and flaws. Many of the observational mistakes were much more obvious by this time, 700 years later. Astronomers in Baghdad, including al-Khwarizmi and al-Kindi, set to work correcting and improving the data in *The Almagest*, by making their own observations, something they were able to do much more effectively in the observatory—the first in the Islamic world—that al-Ma'mun had built in the Shammasiya district of the city. Advances in both their equipment and their methods meant that their data was more accurate than Ptolemy's, enabling them to make significant improvements to his models.

Al-Ma'mun built another observatory just outside Damascus, so that data from the two places could be compared to achieve even greater accuracy. His teams of astronomers designed and built intricate astrolabes, and other specialist equipment included quadrants and sundials that could measure the height of the sun by the length of its shadow.[29] Using these, they corrected and extended *The Almagest*, producing improved versions of Ptolemy's *Handy Tables*, or *Zij*, as they are known in Arabic. These small manuals contained many of the star tables that were dispersed throughout *The Almagest*, making them easily accessible and more useful. The *Zij* revolutionized astronomy and astrology; scholars adapted the data they contained to their own locations and then used them to calculate the positions of the stars and the planets with much greater accuracy than before. They were extremely practical books, so they spread widely throughout North Africa, Spain and Sicily, and the rest of Europe, vital tools in predicting the movements of the stars and planets.

This glimpse into medieval Arab science illustrates the delicate, complex interconnections between astronomy, astrology, philosophy, mathematics and geography. These Baghdadi scholars, with their wide range of interests and expertise, were Renaissance men who prefigured the Renaissance by several centuries. Al-Ma'mun's forays into the desert reveal the rigour and care his scientists took with their work. Their methods of observing and measuring natural phenomena, carefully checking and comparing data, and then developing and testing hypotheses, would be familiar to modern scientists. These principles, together with al-Razi's innovations in medical practice, marked a new era in academic study. And, passed down through the centuries, they form the bedrock of what is now known as "scientific method."

Scholarship continued to burn brightly in Baghdad into the eleventh century, but the Abbasids' grip on power was often tenuous, and the city was plunged into long periods of violence and upheaval. It was finally destroyed in 1258 by a Mongol army, led by Genghis Khan's fearsome grandson, Hulagu (1218–1265), who unceremoniously bundled up al-Mustasim, the last Abbasid caliph, in one of his own ornamental rugs and had him trampled to death by horses. The Abbasid dynasty had come full circle—ending in the brutality with which it had begun. Even amid the destruction, however, strands of knowledge survived. Hulagu, who was apparently "addicted to alchemy and astrology,"[30] had the libraries of Baghdad ransacked and his scholars took the books they were interested in to his observatory on the plateau of Maragheh, in north-west Iran. The rest were destroyed when his army torched the city. Astronomers, notably Nasir al-Din al-Tusi (1201–1274), continued the work of al-Kindi and al-Khwarizmi in the Maragheh observatory, where they built instruments that enabled them to make increasingly accurate observations. The data produced made it possible to challenge and modify Ptolemy's models by introducing, among other things, oscillation to planetary movement.

The Baghdadi golden age was over, but the fame of its scholars had spread outwards like ripples on water. At its peak, the Abbasid caliphate and its court inspired rulers across Persia, Central Asia, North Africa, Spain and the Arabian peninsula to fill their cities with scholars, educate their children, pay for books and build libraries. The fashion for patronizing scholarship was emulated in Cairo, Mosul, Basra, Damascus, Kufa, Aleppo, Tripoli, Bukhara and Shiraz, where great libraries flourished. New generations of scholars rose to prominence, among them Avicenna Ibn Sina al-Biruni, al-Tusi and Ibn al-Haytham, who made their own unique contributions to learning and brought intellectual brilliance to cities like Ghazni, Merv and Cairo.

But the brightest star of all burned far away in the west, in Spain.

The Umayyad family, whom the Abbasids had come so close to annihilating completely, were building a gleaming edifice in southern Spain to compete with the Baghdad of Harun and al-Ma'mun. Córdoba was about to become the new axis around which the world of scholarship revolved, and it is the next stop on our journey.

FOUR

Córdoba

Cordova [*sic*], under the Sultans of the family of Umeyyah, became the tent of Islam, the place of refuge for the learned . . . To it came from all parts of the world students anxious to cultivate poetry, to study the sciences, or to be instructed in divinity or the law; so that it became the meeting-place of the eminent in all matter, the abode of the learned, and the place of resort for the studious.

Her [Córdoba's] necklace is strung with the inestimable pearls collected in the Ocean of language by her orators and poets; her robes are made of the banners of science . . .

—Ahmed ibn Mohammed al-Makkari,
History of the Mohammedan Dynasties in Spain

In the western regions, there shone a fair ornament of the world, an august city, proud as a result of its newfound military might, a city that had been founded by Spanish settlers and was known by the famous name of Córdoba; a wealthy city, renowned for its charms, splendid in all of its resources, overflowing in particular in the seven streams of knowledge, and ever noted for its continual victories.

—Hroswitha of Gandersheim, *Passion of Pelagius*

A palm tree stands in the middle of Rusafa,
Born in the West, far from the land of palms.
I said to it: How like me you are, far away and in exile,
In long separation from family and friends.
You have sprung from soil in which you are a stranger;
And I, like you, am far from home.
May dawn's clouds water you, streaming from the heavens
    in a grateful downpour.

—Emir al-Rahman I

A MAN LIES in the shade of a pomegranate tree. He is old, nearing the end of a long life. His face is lined and ravaged by time; his eyes are pouched and creased, but they can still pierce you with their blackness. Water gurgles down the channel that divides the garden, birds swoop down to drink and all is good with the world. He gives thanks to God that he can lie here, peacefully enjoying his old age. Relaxing back into the brilliant silk cushions, he gazes up at the tree that shades him from the sun's dazzling light. His mind wanders back over his long life, heading eastwards, thousands of parasangs over the deserts, to the land of his birth, to Syria. He sees another pomegranate tree, in another garden, the forebear of this one. Closing his eyes, Abd al-Rahman ibn Mu'awiya ibn Hisham ibn Abd al-Malik ibn Marwan (731–788), the first Emir of al-Ándalus, hears the shouts of his brothers and cousins as they chase each other through shady courtyards, jumping the water runnels and hiding behind fountains. He is back in the old palace of Rusafa, built by his grandfather, the Umayyad Caliph Marwan II, just outside Damascus—a vision of heaven on earth, a vision that has haunted him his whole life, a vision that inspired the garden in which he now lies. He remembers its beauty, the carefree days he spent exploring its shady corners, the tranquillity and happiness. And then his body jolts as he remembers the moment this sunlit world suddenly turned black.

.   .   .

In 750, Rahman was a young man of twenty, one of the Umayyad Caliph Hisham ibn Abd al-Malik's many grandsons, living a life of pleasure and privilege surrounded by his family. He spent his days hunting and hawking with his cousins, flirting with slave girls, teasing his sisters. But, in the spring of that year, his carefree world was turned upside down when the Abbasid tribe seized power and descended on Damascus, intent on murdering every member of the Umayyad family they could find. According to one of the more colorful versions of the story, Rahman escaped with his younger brother and a servant called Badr. They fled, with the sinister black flags of the Abbasid horsemen fluttering in their wake. The next few weeks were spent desperately trying to outrun their pursuers— hiding in forests, begging for shelter in villages and literally running for their lives. Eventually, they reached the bank of the Euphrates and, with the Abbasids hard on their heels, they threw themselves into the water and began to swim. Rahman's brother, exhausted, turned back towards the enemy soldiers, who were shouting from the bank that there was nothing to fear, that they wouldn't harm them. Rahman begged him to keep swimming to the other side, and had to watch helplessly as the Abbasids dragged him from the water and beheaded him on the spot. Safely on the opposite bank, Rahman and Badr ran until they collapsed with exhaustion. They had escaped, for now at least, but they would never set foot in Syria again.

Rahman spent the next four years, if not always on the run, then on the move. He travelled across the deserts of northern Africa, from Egypt through the lands of nomadic Berber tribes. Some were friendly towards him, but the tentacles of the powerful Abbasids, who now held the caliphate, stretched far, and local rulers were easily persuaded that Rahman was a threat. He had many narrow escapes—once hiding under a pile of clothes belonging to a chieftain's wife—and finally ended up in what is now Morocco, homeland of his mother's tribe, the Nafza Berbers. At this point,

Rahman must have allowed himself to breathe a sigh of relief—the first since his speedy departure from Syria. He had managed to put several thousand kilometres between himself and the Abbasids, he had managed to find refuge among his kin, but, most incredible of all, he had managed to stay alive. While he must have felt lucky, he was still a penniless fugitive—a far cry from the all-powerful ruler he grew up thinking he would be. His heritage as a direct descendant of the Prophet made local rulers nervous; no one wants a would-be caliph moving in next door, and, wherever he went, he was the focus of discontent and suspicion. Rahman could not escape his background and nor did he want to. God had spared his life for a reason: the future of the Umayyad dynasty lay in his hands. If he was going to fulfil his destiny, he knew what he needed to do: con-quer new territory and build a new empire. When his initial plan to secure Ifriqiya (Tunisia) failed, he turned northwards and looked over the narrow stretch of water that joins the Mediterranean with the Atlantic, towards Spain, which had been conquered by Arab and Berber tribes forty years earlier.

What did Rahman know about the land that awaited him? The vast "bull-shaped" peninsula, traversed by snow-capped mountains rich in minerals, winding rivers and wild, high plains, was another world from the deserts of Syria he had left behind. Medieval Arab writers are famous for their effusive, flowery style, and their descrip-tions of al-Ándalus, which today we know as Andalusia, are no exception. The natural beauty of this new Muslim country made them breathless with delight. They waxed lyrical about its "gentle hills and fertile plains, sweet and wholesome food . . . great number of useful animals . . . abundance of waters . . . pure and wholesome air . . . slow succession of the seasons of the year."[1] Their reaction is understandable; arriving from the deserts of North Africa and the Middle East, they must have thought this green, temperate land was like heaven on earth.

Muslim invaders weren't the first to discover the wonders of the Iberian peninsula—the Greeks and Phoenicians had built trading cities along the Mediterranean coast centuries before the Romans arrived in 218 BC, taking control of most of the interior as well. With typical efficiency, the Romans divided Hispania (as they called it) into provinces, establishing capitals at Córdoba, Mérida and Tarragona, and began the process of transforming the landscape, harnessing the natural resources and building an entirely new society. With its huge reserves of gold, silver and other metals, mining became big business—Pliny the Elder reckoned that Iberia produced 20,000 Roman pounds of gold a month, equivalent to 6,578 kilograms or 6.5 tonnes. Agriculture was transformed and grain, grapes and olives were exported across the empire. The Romans built a huge network of roads, complete with milestones, shelters and bridges, a network that still forms the basis of the Spanish transportation system. The fishing industry flourished; tuna, mackerel and sardines were salted in their millions and sold all around the Mediterranean, along with vast quantities of garum—a spicy fish sauce used for seasoning food. The Romans ruled Spain for seven centuries, settling and intermarrying with the indigenous Iberian population. In this atmosphere of relative tranquillity, cities grew, culture flourished and the peninsula became famous for its horses, grain and metals.

By the end of the fourth century, the empire was collapsing in on itself and, in early autumn 409, 200,000 members of the Vandal, Suebi and Alan tribes crossed the Pyrenees into Hispania and shattered Roman control of the peninsula. In the turmoil of the following century, however, a different Germanic tribe, the Visigoths, gained the ascendancy and went on to preside over two centuries of general decline. As warriors, the success of their society rested on the need for constant victories, and therefore battles, to keep them happy with booty and lands. They ruled the Iberians as a proportionally tiny elite, without ever really assimilating or creating a new society, as the Romans had done. Constant infighting and an increasingly oppressive attitude to their subjects (especially Iberia's large Jewish community) resulted in stagnation in almost all areas of life. Trade reduced dramatically, there was widespread urban depopulation, and culture shrank to such an extent that some historians have nick-named them the "Invisigoths."

This bleak period in Iberian history came to an abrupt end in 711, when Arab tribesmen from across the Middle East joined forces with Berbers from Morocco and crossed the thirteen kilometres of sea to the southern coast of Iberia. They met the incumbent Visigoth regime's ineffective resistance with decisive military force and gener-ous terms of surrender, so that, within three years, they had taken control of the major cities and the whole southern half of the pen-insula. Their task was helped by the local population, who, sick of Visigothic oppression and depleted by years of famine,* all but wel-comed them as liberators. Settlements were made with local rulers, and land and money were distributed among the Arab and Berber conquerors. The next chapter in the history of Iberia had begun. The following decades were chaotic, as waves of new settlers arrived from Africa and the Middle East. In this melting pot of races, faiths and tribes, different factions struggled to gain the ascendency. A rapid succession of governors appointed by the Emir of Tunisia— who was, in turn, ruled by the caliphate in Damascus (until around 747, still held by the Umayyads)—tried and failed to unite and sta-bilize the region. The only way this volatile cocktail of humanity could be brought under control was by strong, direct leadership. Al-Ándalus desperately needed a powerful ruler and Rahman des-perately needed somewhere to rebuild the Umayyad dynasty. There would have been no doubt in his or his followers' minds that it was all part of God's plan.

---

* Between 707 and 709, almost half the population died.

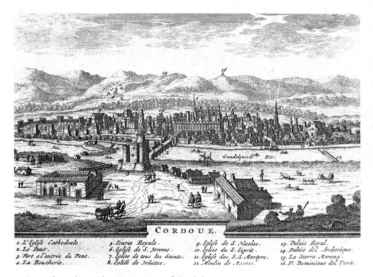

11. An early-eighteenth-century view of Córdoba.

Rahman landed on Iberian soil in 755, in Almuñécar, just to the east of Málaga, and immediately began amassing support. As news of his miraculous survival and escape from the Abbasids spread, people rushed to offer him their fealty, many of them Syrians with Umayyad connections, who had migrated to al-Ándalus in the 740s. Seville fell quickly and peacefully, and, in the spring of 756, Rahman, aged only twenty-five, found himself on the road to Córdoba. When he arrived, the Guadalquivir River was in spate; torrents of churned-up mud and water were swirling under the ancient Roman bridge that led into the city. The current emir, al-Fihri, was waiting for him. In the battle that followed, Rahman defeated al-Fihri, and he rode triumphantly over the river into the city. After years on the run, he had finally found a new home. He moved into the Visigoths' fortified palace, proclaimed himself Emir Rahman I and set about consolidating his control of al-Ándalus. He also began a huge programme of building in Córdoba, inspired by the remains of the glorious Roman civilization that lay all around—ruined temples, derelict baths, decaying civic buildings, statues, mosaics and the skeleton of a sophisticated irrigation system.

Rahman's other great source of inspiration in building his new city was the homeland he had been forced to leave behind. As the poem at the beginning of this chapter shows, he never forgot Syria; it was the guiding light in his creation of a new world in al-Ándalus. In 784, he commissioned a mosque—La Mezquita—on the site of the old Visigothic Church of San Vincente, which, until then, the Christian and Muslim populations had been sharing as a place of worship. The site, just inside the city, by the great bridge and close to the rebuilt and renamed Alcazar (Arabic for "fortress"), had been holy for over a millennium—ever since the Romans had built a temple there. Rahman's mosque was built in the style of the one he had worshipped in as a little boy in Damascus, but, as it was added to by his descendants, it developed its own distinct architectural form: a fusion of Roman, Syrian, Visigothic and Iberian styles. The 800-odd columns (many of them taken from Roman ruins) support two tiers of red and white* striped arches, creating hypnotic patterns and symmetrical vistas. La Mezquita is breathtaking, the most magnificent example of Islamic architecture in the West. But, in 1236, when Córdoba was conquered by Christians from the north, they lost no time in converting it into a church, installing an altar and consecrating the building. A couple of centuries later, the Bishop of Córdoba, Alonso Manrique, decided that, even with the new altar, it did still look alarmingly mosque-like, so he obtained permission to build a cathedral in the middle of it. Walking through La Mezquita today, you notice that the space is resolutely Islamic, the rows of columns fan out in mesmerizing contours as you move through them, but,

---

* White for Umayyads and red for the Prophet (it was allegedly his favourite colour and was associated with blood and life).

when you reach the centre, the ceiling suddenly rises dramatically upwards to pointed arches and elaborate fan vaulting, and you are in a Gothic cathedral, complete with mahogany choir stalls and a crucifix. It is surely one of the strangest buildings in the world, a mosque with a cathedral crouching on top of it, a gigantic stone embodiment of the struggle between the two religions.

Meanwhile, the landscape surrounding the city was also being transformed. Under the Islamic system of tenant farming, farmers paid landowners a percentage of the harvest, rather than a rigid tax as under the feudal Visigothic system. This meant that good yields were in everyone's interest, so people were encouraged to invest in agricultural infrastructure; landowners provided equipment and farmers provided labour. The Arab settlers brought technological expertise in irrigation gathered over centuries of farming in some of the driest places on the planet—it's not surprising that almost every modern Spanish word connected with it comes from Arabic. Across al-Ándalus, people built water wheels (*noria*) to fill irrigation channels, they studied and improved the soil, neatly terraced the slopes into fields and diverted mountain streams into storage cisterns. They restored the network of Roman canals and extended it, increasing the amount of fertile land and improving yields. For every grain of wheat sown, six grains could be expected at harvest—in France, the ratio was only one to three. The relative political stability encouraged farmers to plant valuable, long-term crops, like olives (which can take up to forty years to mature) and vines for producing oil and wine. The final feature of the agricultural boom was the introduction of a cornucopia of new plants, many taken from India and dispersed across the Islamic Empire as it expanded. Monsoon crops, like cotton, sugar cane, bananas, rice, oranges and watermelons, grew well in Iberia, so long as they were properly irrigated. Other exotic delights, such as dates, apricots, aubergines, peas, peaches, saffron and figs, transformed the kitchen gardens and dinner tables

12. Reconstructed Arabic water wheel
near the old Roman bridge over the River
Guadalquivir at Córdoba.

of al-Ándalus, adding to the wide range of existing produce. This abundance of food fed the burgeoning population of Córdoba and filled the emirate's coffers with money—some 100,000 dinars a year, by the end of Rahman I's reign—not bad for a man who had arrived with nothing from the other side of the Mediterranean just a few decades before.

Rahman I's profound love of plants made him the leader of a green revolution. He created a garden palace, just to the north of Córdoba, and named it al-Rusafa, after his grandfather's haven outside Damascus. He sent agents across North Africa and the Middle East, and to Syria, to seek out and bring back plants and seeds—"all kinds of rare and exotic plants and fine trees from every country"[2]—so he could surround himself with the fruit and flora of his childhood and thus build up an incredible collection of plants. One story suggests that he sent a messenger to his sister, who had remained in Syria and survived the Abbasid massacre, to try to persuade her to come and live in al-Ándalus with him. She refused to leave, but sent him a basket of Syrian pomegranates instead—a memento of their happy childhood. The fruit rotted on the long, hot journey, but one of Rahman's courtiers planted the seeds anyway, and their descendants now grow all over Spain.[*] Al-Rusafa was the forerunner of the next generation of Andalusian botanical gardens, where experts studied plants, developed medicines and found ways of acclimatizing exotic species, which were then used in the treatment of disease. The science of horticulture flourished and played a vital role in the development of Andalusian medicine. Later, under Christian rule, the tradition continued with monastic physic gardens, where plants were grown and used to prepare drugs and remedies. But it all began in Rahman's beloved Rusafa, where "these productions of distant regions and various climates failed not to take root, blossom and bear fruit in the royal gardens, whence afterwards they spread all over the country,"[3] transforming the medicine and agriculture of the region.

By the time Rahman I died, in 788, Córdoba was a flourishing centre of commerce and civilization. He had laid the foundations of the great Mezquita that has dominated the city's skyline ever since, and provided a strong political framework for his descendants to build upon. The process of freeing al-Ándalus from Abbasid political influence was accelerated, in 763, when Rahman defeated an army sent from Baghdad, and then had the heads of its leaders labelled, packed in salt and delivered to Caliph al-Mansur. On receiv-

[*] Another version of this story claims that Rahman sent agents to the deserted garden of Rusafa, in Syria, to bring back fruit of the pomegranate trees that grew there.

ing the gruesome delivery, Mansur apparently exclaimed, "God be praised for placing a sea between us!"[4] After that, the Abbasids left al-Ándalus alone. Rahman I had proved to be their nemesis: strong, decisive and a terrifying enemy, but also pragmatic, open-minded and surprisingly sensitive. He continued the policy of religious tolerance practised by Muslims across the empire. The local Iberian population were mainly Christian, but there was also an important Jewish community who had suffered terribly under the Visigoths. The Jews were now allowed religious freedom and were only subject to a few rules, including the *jizya*—the tax levied on non-Muslim citizens. This toleration and cooperation would come to define Andalusian society in the centuries that followed, and would have a profound effect on its scholarship. This enlightened attitude was also extended to the huge (and constantly growing) population of slaves who contributed so significantly to the prosperity and success of al-Ándalus. Many were Slavs, captured by Frankish warriors during the wars on their eastern frontier and taken to Spain to be sold into servitude. Those seized as children grew up in the traditions of their new land—huge numbers converted to Islam and were rewarded with their freedom.

Córdoba's development into a major centre of power was slow in comparison to its main rival, Abbasid Baghdad. By the time Rahman II (great-grandson of Rahman I) became emir, Baghdad's golden age of scholarship was well under way. The relationship between the two cities was complex; the bitterness and rivalry between the two ruling dynasties was legendary and the Umayyads' fervent desire to become independent from Abbasid power would culminate in their proclaiming a rival caliphate in the early tenth century. While the two were gradually separating politically, the opposite was true of culture, administration and commerce. As trade networks grew across the Islamic Empire, the traffic of goods flowing between Córdoba and Baghdad (and all the places in between) became a flood. In the ninth century, Baghdad was the cultural centre of the Dar al-Islam, so Córdoba, teetering right on the very edge of the Arab world—and, indeed, the known world—looked to it for inspiration in everything. The structure of the Umayyad state also owed much to that of medieval Iraq: the postal service, the system of duties imposed on imports and exports, and the currency were all borrowed ideas. Cultural transaction between the two states was embodied by the singular figure of Ziryab, the legendary Persian singer, who left the Abbasid court in Baghdad and arrived in Córdoba in 822, where he spent the rest of his life teaching the Andalusians how to live in style. The esteem with which they held Eastern culture was made clear from the moment of Ziryab's arrival; Rahman III "not only rode forth himself to receive and welcome him, but entertained him for several months in his own palace, and made him considerable presents."[5] Along with his extraordinary voice, Ziryab brought to Córdoba the full wonder and sophistication of the Abbasid court—he is credited with bringing al-Ándalus into the ninth century, so to speak, by introducing a huge range of fashionable innovations, including toothpaste, courses at meals, asparagus, cutlery, tablecloths, hairstyles, clothing and new instruments and styles of music. He became a close confidant of the Emir, who fell under the spell of this charming, sophisticated man. Ziryab was also learned, and he encouraged the study of astronomy and geography at the Córdoban court. He became a cultural icon, the man who showed the Andalusians what they could learn from the East and gave them the confidence to develop their own ideas.

As Andalusian culture developed over the course of the ninth century, the idea that knowledge should be sought through travel began to take root, encouraged by the intellectual brilliance of the East. Young men began to set off into the unknown to "find themselves," learning from the best thinkers of the time and suffering the inevitable deprivations and terrors of early medieval travel.* A *rihla*, as these journeys were called, was primarily a search for religious enlightenment, but, in reality, it often involved acquiring secular, scientific knowledge as well—at this point, there was no real division between the two. This was partly down to the Islamic system of education, in which masters were venerated and their teaching sought by anyone wishing to learn and open their minds. Trade had opened up the Muslim Empire, rulers built and repaired roads connecting places, creating an infrastructure through which people and goods could move relatively easily. Scholars travelled with merchants' caravans and spent the long desert evenings together around the caravanserai fires. Merchants, naturally cosmopolitan and open-minded, were often scholars in their own right, using their commercial activities to acquire books and bring them back to al-Ándalus to be copied and sold. By now, too, there were specialized book and paper merchants, responsible for producing, trading and moving texts between the great book souks of Cairo, Fez, Baghdad, Timbuktu and Córdoba. These were the major channels through which the rivers of knowledge flowed around the Islamic Empire. Merchants and scholars were joined in the caravans by another type of traveller—pilgrims. One of the fundamental pillars of Islam was the need to perform the hajj, the pilgrimage to Mecca, at least once in their lives, which made travel an essential part of Muslim life.

Hundreds of scholars and pilgrims made the long, arduous journey east, to Arabia and Iraq, bringing back new ideas and books when they returned to al-Ándalus. In the 820s, Mu'tazili theology was introduced to Córdoba by scholars who had encountered it in Iraq. As its teachings were disseminated, Andalusian intellectuals

---

* During this period, debate raged as to whether the book could replace the master as the primary means of attaining knowledge.

opened up to the idea that Greek logic could be used as a framework for investigating philosophical questions, within an Islamic context. This was the beginning of a tradition that produced various scholars over the next century, including the first Andalusian philosopher, Muhammad b. Masarra al-Jabali (883–931). Al-Jabali's father had travelled east in the mid 850s, learned Mu'tazili ideas in Basra, and brought books on the subject back with him. At that time, al-Ándalus was dominated by religious conservatives, so these early Mu'tazili scholars had to be careful not to attract too much attention from the authorities—some were persecuted and their books were burned. Even though it was underground at first, Mu'tazilism helped to bring classical learning to al-Ándalus, coming into its own during the following century, under the enlightened rule of Rahman III and al-Hakam II—just as it had in al-Ma'mun's Baghdad.

Rahman II, who ruled from 822 to 852, opened up trade routes in the Mediterranean by making alliances with the Byzantines in Constantinople. This increased opportunities for the trade of Andalusian produce, minerals and textiles, created huge wealth and connected the peninsula with the wider world. Rahman was also a generous patron of scholarship and did all he could to stimulate intellectual activity in Córdoba. By the mid ninth century, Arabic culture was flourishing—as is clear from the complaints of the Christian scholar Paul Alvarus, who bemoaned how young Christians had fallen in love with the Arabic language and its poetry: "All talented young Christians read and study with enthusiasm Arab books; they gather immense libraries at great expense . . . they have forgotten their own language."[6] That language was, of course, Latin, slowly suffocating from a lack of ideas and religious atrophy while Arabic triumphed—exotic, poetic, the language of the future, the language of science. No wonder young Christians were keen to learn it and participate in the exciting new culture that had transformed their city. They came

13. View of the Roman bridge over the River Guadalquivir with Córdoba on the left bank. The cathedral roof is clearly visible on top of the Mezquita, while the edge of the Calahorra Tower can be seen on the right of the picture at the opposite end of the bridge.

to be known as Mozarabs—Arabized Christians—and grew into a large and influential community, spread across Andalusia.

While conservative Christians like Alvarus, now second-class citizens, felt marginalized and threatened in Córdoba's multicultural, Muslim-dominated society, the opposite was true of the city's large Sephardic Jewish community. Having endured the Visigoths' persecution, they thrived under the new, comparatively broad-minded regime, which allowed them to build synagogues and live peacefully in the Jewish quarter of the city, just north of the Alcazar. Unlike the Christians who had lost their dominant position to the Arab settlers and Islam, the Jews were used to retaining their own language, faith and society alongside that of the country in which they lived. Young Jewish men also embraced Arabic language and culture, and the tolerance of the Umayyad society allowed them to succeed in many areas of public life and rise as high as their talents took them. Jewish scholars played a fundamental role in the transmission of science in later centuries, while their community remained a pillar of civic life in Córdoba. They were especially prolific in medicine, making up 50 per cent of doctors in Spain, while only representing 10 per cent of the wider population.

It was around this time that scientific ideas started to arrive from the East in significant numbers. As we saw earlier, by the beginning of the ninth century, the great period of translation was under way in Baghdad and the Islamic book trade was booming. Rahman II's successor, Muhammad I, who reigned from 852 to 886, created a royal library that was the largest collection of the time, and the Andalusian elite spent thousands of dinars following his example. The book markets were crowded with wealthy men on the look-out for the best volumes to fill their library shelves. However, this wasn't good news for scholars—one complained that, when a book he had been seeking for months finally turned up in an auction, he found himself caught in a bidding war. The price went so high that he had to give up and lost the book; his disappointment turned to anger when the man who outbid him admitted that he had no idea what it was about, he was simply, "anxious to complete a library which I am forming, which will give me repute amongst the chiefs of the city."[7] The age-old squabble between wealthy dilettantes and penniless scholars had reached al-Ándalus. There isn't much specific information in the sources about how individual books arrived there, but a poet and jurist called Abbas ibn Nasih, who lived in Egypt and Iraq for many years, is named as having brought books from the East to Rahman II in Córdoba. He is just one example, but there must have been many more like him—men who either presented texts as gifts when they reached al-Ándalus, or sold them to scholars and collectors. Another man who definitely brought books back from Iraq was Abbas ibn Firnas, who also happens to be the first great figure in the Andalusian pantheon of scholarship. Born in Ronda in 810, this "Leonardo da Vinci of Islamic Spain"[8] had an extraordinary range of scholarly interests, and was employed as a court astrologer by Rahman II. It isn't easy to ascertain the actual facts of his life, but he is said to have travelled to Baghdad to study, before returning to al-Ándalus, where he taught mathematics and music, wrote poetry, invented an ingenious method of cutting rock crystal,* and designed and constructed a water clock, an armillary sphere (used in astronomy) and a planetarium. But he is best known for attempting to fly by covering himself in feathers and jumping off a tower (or cliff, depending on the story), holding on to specially designed wings. By some miracle, he survived, despite being in his sixties, concluding that he had not realized how important birds' tails were in the landing process.

In spite of his foolhardy experiments, Ibn Firnas lived to be an old man. He helped initiate a tradition of scholarship that would reach its apogee in the following century when Rahman III was on the throne. Abd-al-Rahman III was born in 891. He was the grandson of the seventh emir, Abdullah, who controversially passed over all four of his sons in naming his young grandson as his successor. Rahman's mother was a Christian slave, and his grandmother was a Christian princess, the daughter of the King of Pamplona, so the new emir had a mixed ethnic and religious background, and blue eyes and fair hair, which he apparently dyed black to make himself look more Arab. Abdullah died leaving al-Ándalus in chaos, riven by internal disputes and rebellions, threatened from the north by the Christian King of Asturias and from the south by the Fatimids of North Africa. Contemporary political commentators were prob-

---

* Craftsmen created beautiful jugs and glasses made from pieces of rock crystal, which they painstakingly hollowed out and decorated with intricate etched patterns.

ably not filled with optimism when the twenty-one-year-old Rahman ascended the throne; they could not have known that he would turn out to be the most important, successful leader in the history of Islamic Spain. Well educated and erudite, Rahman spoke several languages fluently and was an enthusiastic patron of scholarship. He founded a university in the Mezquita and encouraged scholars to work on the scientific texts that had been brought from the East. Al-Khwarizmi's work on Hindu arithmetic and his astronomical tables (*Zij*), which were probably brought to al-Ándalus by Abbas ibn Firnas in the mid ninth century, were very influential, informing Andalusian astronomy. Under Rahman III's patronage, al-Majriti (whose name suggests he was born in Madrid) adapted the *Zij* to the longitudinal coordinates of Córdoba so that they could be used to predict the movements of the stars, ascertain the direction of Mecca and work out the correct time of day for prayers. Astronomy was a vital tool of power, and Rahman understood the importance of filling his court with scholars devoted to studying the stars and working out how to predict their movements, and—correspondingly—foretell the future.

Al-Majriti was a leading member of the circle of scholars at Rahman's court. They made regular observations, working together to improve the accuracy of their tables and correct astronomical theory. Al-Majriti was a great teacher and mentor to the next generation of scientists, and a school grew up around him. He was thus an extremely influential figure in the development of astronomy and mathematics in al-Ándalus, and as his students moved to other cities, they took his ideas with them. Al-Majriti was "very fond of studying and understanding the book of Ptolemy known as Almagest,"[9] and reportedly also made a translation of Ptolemy's *Planisphaerium* ("*Star Chart*"), which has not survived in Arabic, only in the Latin translation that was made in Toledo in the twelfth century. Al-Majriti taught his students how to use astronomical instruments properly, and how to make their own. His pupil, al-Saffar, continued al-Majriti's work on astrolabes, and the treatise al-Saffar later wrote was so important that it was still being used by astronomers in the fifteenth century. Al-Majriti, al-Saffar and another of his students called al-Samh all made new versions of al-Khwarizmi's *Zij*, adapted to the geographical position of Córdoba. Al-Samh also wrote a book explaining the geometry in Euclid's *Elements*, and two treatises on the astrolabe.

As a result, al-Ándalus produced a rich tradition of astronomical study, grounded in rigorous observations taken over long periods of time. Andalusian astronomers adapted the latest theories from the East to their own needs and location, they went back to Ptolemy to study, challenge and correct his work, and they made use of ideas from Indian mathematics and from Euclid's *Elements* to produce their own contributions to the gradual process of assessment and improvement that drives scientific research. Their progress was enhanced by the high standards of local craftsmanship, especially in metalwork, which they exploited to make ever more accurate, useful instruments. It is no coincidence that al-Saffar's father was a brass worker, and this fertile communion of art and science produced some astonishingly beautiful objects.

Rahman's court drew the most brilliant and ambitious young men from across the peninsula. None was more brilliant or ambitious than a young Jewish man from Jaén called Hasdai ibn Shaprut, who was initially employed as a court physician. Like his master, Rahman, Hasdai was a talented scholar, fluent in several languages (including Latin, which was otherwise understood only by a small handful of Christian priests), charming, sophisticated, clever—everything a ruler could wish for in an advisor. Before long, Hasdai had become indispensable to Rahman, who put him in charge of customs and imports in Córdoba, a position he used to transform the state of the treasury. Hasdai's proficiency for languages, his intelligence and his prestige in the international community of Judaism made him a perfect diplomat. He communicated with the court of the Holy Roman Emperor, Otto I, in Frankfurt and the court of the Byzantine Emperor in Constantinople and when ambassadors arrived in Córdoba, it was Hasdai who received them. John of Gorze, a monk who was part of a delegation sent from Frankfurt, said he had "never seen a man of such subtle intellect."[10] Hasdai's talents seem to have known no bounds; he even managed to cure the King of Leon, Sancho the Fat, of his obesity. Hasdai was joined in Rahman's inner circle by the Christian (Mozarabic) Bishop of Elvira, Recemund, whose religion and background made him eminently suitable for the diplomatic missions he made to Constantinople and Frankfurt. Recent histories of al-Ándalus have sought to challenge the idea that it was a place of particular religious tolerance, but Recemund and Hasdai's importance in Rahman's government demonstrate that, at the upper level of society, open-mindedness was clearly dominant. Moreover, the fact that they weren't Muslim was crucial to the roles they played.

By 929, Rahman had stabilized al-Ándalus, using a combination of brute force and skilful negotiation. Córdoba was, by this point, a large, wealthy city. Its scrupulously clean streets were lit with lanterns at night, the scent of delicious food and fragrant spices filled its bazaars, while water flowed through the irrigation system into countless fountains in the shady courtyards of Córdoban houses. In the city's workshops, craftsmen created the most beautiful tooled leather, intricate filigree jewellery, lavish fabrics and the famous copper green and manganese blue pottery—luxurious goods that were sold all over the Mediterranean and the Middle East, creating enormous fortunes for the city's artisans and merchants. Wealthy Córdobans built incredible palaces in the lush valley of the Guadalquivir, inspired by the remains of the Roman villas they found there, complete with gardens, orchards, baths and libraries. Even the most fervently anti-Islamic writers were dazzled, describing the city as "a

Raphael's *School of Athens*, with Plato and Aristotle in the centre. Euclid (or Archimedes) is bending over a set of compasses in the front right, while Ptolemy stands beside him with his back to us wearing a crown. Averroes leans forward on the left-hand side, in a green robe and a turban.

The Doric columns of the original Greek temple of Athena, built in the fifth century BC, incorporated into the walls of the seventh-century cathedral in Syracuse.

3. A thirteenth-century Arabic depiction of Aristotle teaching a pupil, quite possibly Alexander the Great.

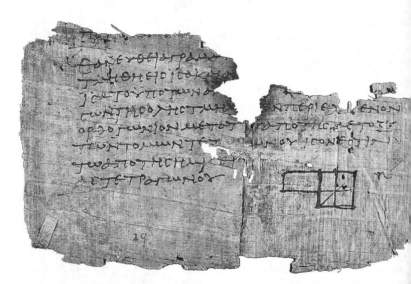

4. The oldest surviving fragment of Euclid's *Elements*, written in about AD 100 and complete with a diagram, found among hundreds of thousands of other flakes of papyrus in an ancient rubbish tip in Egypt. So far, around 5,000 have been pieced together and deciphered, an estimated 1–2 per cent of the total found.

5. Pages from the copy of *The Elements* that Peyrard discovered among the Vatican manuscripts, containing an older version of the text, closer to Euclid's original and not seen for a millennium.

Pages from an early copy of Ptolemy's *Almagest* in Greek.

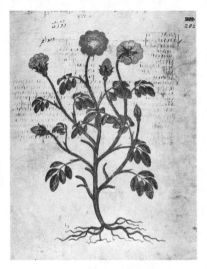

7. Delicate depiction of a rose from the most beautiful edition of Dioscorides' *De materia medica* in existence, created for the Byzantine princess Anicia Juliana in the early sixth century.

8. A detailed depiction of an Arabic library. This is one of a lavish series of illustrations produced in the thirteenth century in a new edition of an earlier text, the *Maqamat*, by the Basran writer al-Hariri, showing life in the medieval Middle East in astonishing detail.

9. Another image from the *Maqamat* showing a doctor treating a patient with "cupping," the ancient practice of creating a localized vacuum on the skin using heat or cold.

10. A sixteenth-century image of astronomers in the Galata Observatory, Istanbul, using a wide array of instruments: a globe, quadrants, sand timers and an astrolabe.

11. A gold Abbasid dinar, struck during the reign of Caliph al-Mamun.

12. Inside the Mezquita. The red-and-white-striped arches create a uniform, rhythmical pattern across the vast horizontal space.

13. A copy of *The Almagest* in Arabic made at the end of the fourteenth century.

14. A stunning example of Toledan craftsmanship, this intricately carved sword once belonged to Boabdil, Muhammad XII of Granada.

15. A saphaea astrolabe, with universal projection, based on the design by al-Zarqali. This one was made in North Africa, probably in the thirteenth century.

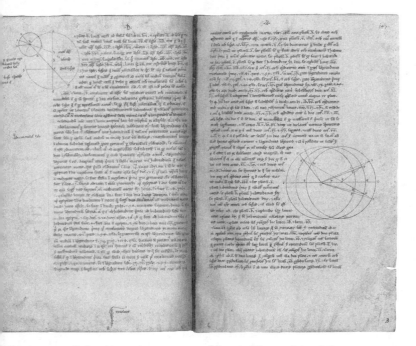

16. An exceptionally fine thirteenth-century copy of Gerard of Cremona's translation of *The Almagest*.

17. A fourteenth-century Latin copy of *The Almagest*, showing animals using astronomical instruments.

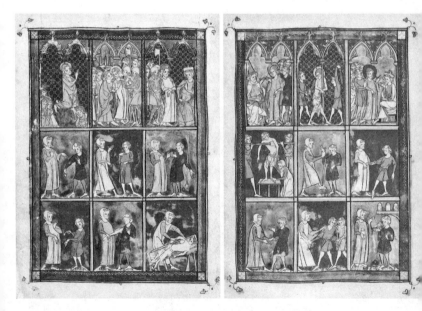

18. Lavishly painted pages from Roger of Salerno's twelfth-century *Chirurgia*. The top rows depict scenes from the life of Christ; the illustrations underneath show doctors consulting patients with an array of complaints and wounds—one man has a huge lance through his forearm—and some of the remedies they prescribed.

fair ornament of the world."¹¹ The moment had come for Córdoba to step out from the shadows of her old rival, Baghdad, for the final twist in the great dynastic feud between the two families. In 929, Rahman III declared a rival caliphate, cutting all ties with Baghdad and exacting the ultimate revenge on the Abbasids, whose empire was disintegrating around them.

But Rahman knew that simply declaring himself caliph wasn't enough. He needed to build himself a magnificent stage to reflect his new power and status. So, in 936, he travelled a few kilometres out of Córdoba, towards the north-east, up into the Sierra Morena, the range of hills that frame the valley. He found a place halfway up the slope of Jabal al-Arus, or Bride Hill, with a panoramic view across the Guadalquivir plain, and began building the magnificent palace-city that he named Madinat al-Zahra. Thousands of work-men, many of them slaves,* worked feverishly to create an entirely self-sufficient urban complex (over a kilometre square), with its own baths, workshops, mosques, bakeries, barracks and, of course, palatial apartments for the Caliph, his family and his government. Mountain streams were diverted, via a restored Roman aqueduct, into huge cisterns, so that every corner of the city had running water, while the site was divided horizontally into three terraces, with pal-aces on the top level. The quarries of Iberia and North Africa went into overdrive to produce tonnes of the best marble: rose and green from Carthage, white from Tarragona. Ancient buildings as far away as Narbonne and Rome were plundered to amass the 4,000 columns that held the city up. In the Hall of the Caliphs, the walls glowed with translucent marble and gold, reflecting off an enor-mous pearl, sent as a gift from Constantinople, that was suspended from the centre of the ceiling. On the floor, a basin of mercury quiv-

---

* The number of Rahman III's slaves is variously reported as 3,750, 6,087 and 13,750. They were trafficked from across Europe and the Black Sea.

ered, terrifying and amazing visitors by sending sunbeams flashing around the chamber.

With the magnificence of the Madinat al-Zahra as his backdrop, Rahman III could take his place on the world stage alongside the other great leaders of the medieval world. His glittering audience chambers thronged with ambassadors from the Franks, the Lombards, the Sardinians, the Byzantine Empire and the Christian kingdoms of northern Spain, while his representative, Hasdai, charmed them all. This flurry of diplomatic activity had many positive outcomes. Rahman's position as caliph was confirmed, al-Ándalus became a major player in international politics and enjoyed powerful alliances, and, as Hasdai proudly pointed out, "Kings of the earth, to whom his [Abd al-Rahman's] magnificence and power are known, bring gifts to him, conciliating his favour by costly presents, such as the King of the Franks, the King of the Gebalim, who are Germans, the King of Constantinople, and others. All their gifts pass through my hands, and I am charged with making gifts in return."[12] Keen to demonstrate their own wealth and cement the alliance against the Abbasids, the Byzantines were especially generous. The Madinat al-Zahra glittered with enormous jewels, marble columns and ornate basins sent from Constantinople. In 949, the Emperor Constantine VII, having heard of Rahman's interest in science and learning, sent something even more precious—a book. Written by Dioscorides in the first century AD, *De materia medica* was a huge, five-volume pharmacopeia, describing 600 plants and their medicinal properties. This copy had been illustrated by the most talented scribal artists in Constantinople and was filled with beautiful pictures of plants, minerals and animals—not mere decoration, but vital tools in identification. It was an extremely important text—and had been for centuries, used by generations of doctors—and a major source for Galen's works on medical botany. Andalusian scholars had already had access to a poor Arabic translation of the text that Hunayn ibn Ishaq had worked on in Baghdad, but many of the plant names had been simply transliterated into Arabic rather than properly identified (which could have chilling consequences when they were used to make medicines); many of the plants didn't actually grow in Iraq, so were unknown to the early translators.

The new translation of *De materia medica* was a turning point in the development of Andalusian medicine, allowing it to surpass the achievements of the East and establish an independent medical tradition in Spain. But, before this could happen, the Córdobans had to have it translated from Greek into Arabic, and there were no Greek speakers in al-Ándalus. Hasdai ibn Shaprut immediately wrote to the Emperor in Constantinople, asking for help, and, a couple of years later, a Byzantine monk called Nicholas arrived to join the team of translators working on the text. Nicholas taught Greek to the Latin-speaking Mozarabs, so that they could interpret between him and the Arab scholars, and, in this way, the great work was gradually translated into Arabic; almost all the plants were identified locally. This multilingual, multiracial enterprise was overseen by Hasdai ibn Shaprut, who encouraged and supported the team of scholars, and doubtless contributed his own considerable expertise when needed. Just as it had been in antiquity, *De materia medica* was one of the most influential medical texts of the Middle Ages. It remained the primary authority on its subject for 1,500 years, copied in Greek, Latin and Arabic, and widely transmitted. During the sixteenth and seventeenth centuries, it was translated into French, Italian, German, English and Spanish, and formed the basis of many Renaissance herbals. The illustrations gave birth to the genre of botanical drawing that straddles the worlds of art and science to this day. The earliest surviving illustrated manuscript was made in the fifth century in Constantinople for the Byzantine princess Anicia Juliana, and is now in a library in Vienna. The Arabic *De materia medica* was a seminal text in the development of Andalusian botany and pharmacology; it was the starting point for a long tradition of studying and describing "simples," remedies made from a single plant, and by the mid thirteenth century, Andalusian scholars had listed over 3,000 of them. They carried out this work in botanical gardens, where they propagated, grew and studied plants and their medicinal properties, developing treatments and laying the foundations of modern pharmacology, continuing the tradition first started by Rahman I in his garden at Rufasa.

Dioscorides' great work joined those of Galen, which were already widely studied in Iberia in the Arabic translations made by Hunayn ibn Ishaq and his circle in Baghdad. One young scholar, who, like so many of his countrymen, combined an interest in science with a talent for writing poetry, elegantly summed up in this verse:

> When I have no guests or companions,
> I entertain Hippocrates and Galen.
> I take their books as a remedy for my loneliness.
> They are the cure for every wound I treat.[13]

Many physicians travelled east to learn from the great masters of Iraq. Two brothers, called Umar and Ahmad, left al-Ándalus and spent ten years in Baghdad, studying Galen with Thabit ibn Sinan, son of the legendary Sabian scholar Thabit ibn Qurra, and ophthalmology with a specialist eye doctor. When they returned home in 962, they were employed as court physicians and became famous for their cures and, in particular, their ability to treat eye complaints. Around the same time, another Andalusian scholar, called al-Jabali, was living in Basra, learning medicine and logic. From there, he moved to Egypt, where he worked as the director of a hospital

before coming back to Iberia in 971.* He built up an impressive reputation as a doctor, renowned for his extensive knowledge and profound understanding of medicine. The fact that scholars had studied
in the East would, no doubt, have given them kudos once they got
back home, but the high regard in which these three doctors were
held must also have been based on their medical practices and ability to cure people. Their contribution to the process of transmission
was vital; they brought the latest medical ideas and translations of
ancient texts back with them and disseminated those texts to the
next generation of doctors in al-Ándalus.

One man dominated that generation. Abu al-Qasim Khalaf
ibn Abbas al-Zahrawi (Latinized to Albucasis, 936–1013) was a true
giant in the history of medicine. We do not know whether he travelled east in search of knowledge, but he was taught by al-Majriti,
and employed as court physician by al-Hakam II, Rahman III's son
and successor, so he spent a large part of his life at the palace city
of Madinat al-Zahra—hence his name. Towards the end of his life,
Zahrawi wrote a vast compendium entitled *Kitab al-Tasrif* (usually
known as *The Method of Medicine*),† which went on to be a cornerstone of medical practice in the later Middle Ages, taking its place
alongside al-Razi's *al-Hawi* and Ibn Sina's *Canon* on doctors' bookshelves throughout Europe.‡ A comprehensive guide to medicine, the
Kitab al-Tasrif was made up of thirty treatises dealing with diseases,
symptoms and treatment, diet and the preparation of simple and

---

* There is no detailed evidence about hospitals in Córdoba, but al-Makkari claims
that there were fifty—probably an exaggeration. It is plausible that al-Jabali
brought innovations and practices back with him from Egypt and Basra.

† Its splendid full title is *The Arrangement of Medical Knowledge for One Who Is
Not Able to Compile a Book for Himself.*

‡ Later physicians were in awe of Zahrawi's genius; some ranked him alongside
Hippocrates and Galen for his contribution to medical science.

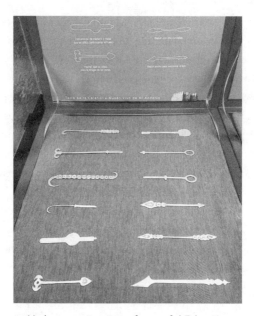

14. Modern reconstructions of some of al-Zahrawi's intricate surgical instruments, displayed in the Calahorra Tower Museum. The implement at the front on the left is described as "an axe for use in surgery of the veins."

compound drugs, plasters, and cures and ointments in the *Materia medica* tradition. The final section, which makes up around one fifth of the whole, was devoted to surgery, and it is for this that Zahrawi is most famous.

In his introduction, Zahrawi bemoans the fact that "the skilled practitioner of operative surgery is totally lacking in our land and time," before going on to emphasize the necessity of the practitioner "to be trained in anatomy as Galen has described it, so that he be fully acquainted with the uses, forms, and temperament of the limbs; and also how they are jointed, and how they may be separated; that he should understand fully also the bones, tendons, and muscles, their numbers and their attachments; and also the blood vessels, both arteries and veins."[14] He was also influenced by Arabic physicians, notably al-Razi, and by the late antique encyclopaedist Paul of Aegina, whose work on surgery was an important source of information.

Zahrawi's practical approach was underpinned by the design of new instruments, many of which have been used ever since: the forceps, speculum, bone saw, lithotomy scalpel, concealed knife, surgical hook, tonsil guillotine, spoon and rod, needle and syringe are just some of them. Not only this, he also produced detailed diagrams and explanations on how to make and use these instruments, a first in the Muslim world. Reconstructions of them are on display in the Calahorra Tower Museum in Córdoba. Laid out on bright red cushions, they could easily be mistaken for beautiful pieces of jewellery, but the labels underneath leave you in no doubt as to their functions: "axe used for surgery of the veins," reads one. Zahrawi's approach to medicine was innovative and pragmatic; he pioneered anaesthetics by giving his patients sponges soaked in opium and cannabis to inhale, and was the first person to use catgut for internal stitching—a natural substance that dissolves easily inside the human body without causing infection, used by doctors ever since. He developed treatments for psychological complaints, including one that was based on opium, which he called "the bringer of joy and gladness, because it relaxes the soul, dispels bad thoughts and worries, moderates temperaments, and is useful against melancholy."[15] His emphasis on anatomy and physiology influenced generations of physicians, as did his ideas about ethics, hygiene, education and diet. He was also interested in the psychiatry and education of children, which he discussed at length in the *Kitab al-Tasrif*. However, it was his book on surgery that had the most profound influence; it was translated into Latin in Toledo and, from there, disseminated throughout Europe.

Zahrawi was lucky that the early part of his career coincided with the apogee of the Córdoban golden age under Rahman III and his son, al-Hakam II. Al-Hakam was even more enthusiastic about learning than his father, and, when still a young man, "he began his effort to support the sciences and befriend the scientists. He brought from Baghdad, from Egypt, and from other eastern countries the best of their scientific works and their most valuable publications whether new or old."[16] Al-Hakam used his time as crown prince well, building a vast network of scholarly contacts and sending agents across the Dar al-Islam and beyond to buy, copy, borrow or steal books—ancient and modern—for his library, back in Córdoba, no matter what the cost. According to Sa'id al-Andalusi, who wrote a detailed survey of the history of science in the eleventh century, "His collection became equal to what the Banu Abbas [the Banu Musa brothers of Baghdad] were able to put together over a much longer period. This was possible only because of his great love for science, his eagerness to acquire the virtue associated with it, and his desire to imitate sage kings."[17] In doing so, he founded twenty-seven schools for poor children, and supported the university founded by his father in the Mezquita, making it famous by granting generous endowments, inviting Eastern masters to come and teach in the city and spending thousands of dinars installing water pipes and mosaics from Constantinople. When al-Hakam ascended the throne in 961, he combined the three principal royal libraries—the palace library, his brother Muhammad's collection and his own—into one, where he allegedly employed 500 people. When the librarian, Talid, catalogued the 400,000 or so books it contained, he filled 404 volumes with the titles alone. Had they survived, they would have given us a detailed picture of the Andalusian intellectual world; as it is, we have to try to reconstruct the library's contents from the surviving works the scholars themselves wrote, and from other sources, such as they are. Al-Hakam II's considerable influence in the bibliographic networks across the Dar al-Islam was partly down to his own personal reputation as a scholar. According to one Andalusian historian, he had read and annotated most of the books in his library; even allowing for exaggeration, this was no dilettante collector. As well as book agents, he employed foreign scholars in Egypt and Baghdad to collect books for him, and, in one case at least, was able to persuade a writer in Iraq to send him a manuscript of his new book before anyone else saw it.

Córdoba's reputation as a great centre of learning drew scholars from far and wide, especially in the fields of medicine, astronomy, religious law, grammar and poetry. Córdobans were so famous for their love of the written word, it was said that, "When a learned man dies at Seville, and his heirs wish to sell his library, they generally send it to Cordova to be disposed of."[18] The grand Ibn Futays family had a library in which they employed no fewer than six copyists and a famous writer as librarian; another notable private collection was created by a poet called Ayesha. According to one source, "To such an extent did this rage for collection increase, that any man in power, or holding a situation under government, considered himself obliged to have a library of his own, and would spare no trouble or expense in collecting books, merely in order that people might say,—such a one has a very fine library, or he possesses a unique copy of such a book, or he has a copy of such a work in the hand writing of such a one."[19] All these books had to be copied, so it is no surprise that the scribal industry boomed in tenth-century Córdoba. Hundreds of people were employed to produce the estimated 70,000 to 80,000 books created in the city each year, in the largest book market in the Western world.

Córdoba's star burned bright, but it burned out very quickly. When al-Hakam died in 976, stability and intellectual freedom died with him. He left his eleven-year-old son, al-Hisham, on the throne, and the vizier al-Mansur saw his chance to seize power. Al-Mansur's megalomania knew no bounds. He set about destroying and plundering al-Zahra to build a new palace for himself on the other side of Córdoba. The vast, magical city on the hillside above Córdoba gradually sank into the dust, used as a base by rival factions as the caliphate collapsed, its treasures ransacked—pipes, columns, carvings, doors and capitals all sold off for the highest price. In spite of this, the ruins of Madinat al-Zahra are powerfully evocative. The grand arches of the audience chambers, the courtyard pools and the dramatic setting impress thousands of tourists today, but it is the tiny details that bring the city back to life—the circular holes where the doorposts were fitted, the up-cycled Roman sarcophagus used as a water trough in the palace stables, the broken pieces of lead piping that were part of the irrigation system.

Al-Mansur also turned his wrath on Córdoba itself. Under the influence of conservative theologians, he ransacked the city's great libraries, looking for all books "dealing with the ancient science of logic, astronomy, and other fields, saving only the books on medicine and mathematics. The books that dealt with language, grammar, poetry, history, medicine, traditions, hadith, and other similar sciences that were permitted in al-Ándalus were preserved." He ordered the rest to be destroyed and "only very few were saved; the rest were either burned or thrown in the wells of the palace and covered with dirt and rocks." His justification was that, "these sciences were not known to their ancestors and were loathed by their past leaders."[20] They were heretical and anyone pursuing them was guilty of not following Islamic law. How they chose which texts to save and which to destroy is a mystery; it must have been difficult to decide whether texts like Ptolemy's *Almagest* were mathematical or astronomical. Córdoba was sacked several times in the chaotic decades that followed, and its libraries ruined. The books that did survive were "scattered all over al-Ándalus,"[21] taken by scholars fleeing to other cities, including Seville, Granada, Zaragoza and Toledo. These small Muslim principalities, known as "taifa" states, gained their independence in the early eleventh century and prospered as Córdoba declined. Their rulers, inspired by the Umayyads, strove to make their courts great centres of learning, with libraries and brilliant scholars. How successful they were depended on a combination of luck, money and their own intellectual interests. Zaragoza was known for its philosophers and, in the late eleventh century, it was ruled by Yusuf al-Mutamin ibn Hud, probably the most gifted mathematician in the whole of Muslim Spain, and certainly the most original. It is clear from his writings that his library was well stocked with every important mathematical text available in the tenth century, with Euclid's *Elements* at its core.

Córdoba never regained the glory it enjoyed under the Umayyads, but it remained a centre of learning and books. In the twelfth century, two of the world's greatest thinkers were born in the city: Maimonides (*c.*1135–1204), the Jewish philosopher whose writings influenced scholars across the Middle East and Europe, and ibn Rushd (Latinized to Averroes, 1126–1198), who is known as the founding father of secular thought in Western Europe because of his widely diffused commentaries on Aristotelian philosophy—the only non-Greek depicted by Raphael in his *School of Athens*. As Spain was gradually reconquered by the Christian kingdoms of the north, the books and ideas of these two scholars form a bridge between early medieval Arab science and the Latin culture of the later Middle Ages. Very few survive in the Arabic versions. We know that copies of Euclid's *Elements*, Ptolemy's *Almagest* and the Galenic corpus came to Córdoba from the East, were copied and distributed across the country and pored over by scholars. Some were taken to Toledo, where they were translated into Latin. Many others had a different fate. In 1492, the last Muslim stronghold, the beautiful city of Granada, fell to the Christians. The terms agreed were generous and enlightened: Spanish Muslims would be allowed to live in peace, practise their religion and follow their own customs. But these hopeful beginnings were soon buried under a wave of intolerance and persecution. There was no place for alien cultures or religions in the Spain of Ferdinand and Isabella; they expelled thousands of Jews, they oppressed and exiled Muslims, and began the process of destroying 700 years of Muslim civilization. The culmination came in 1499, when the fanatical cleric Cardinal Ximénez de Cisneros arrived in Granada intent on converting the population and removing any vestiges of Islamic culture. He took the contents of the city's libraries and built an enormous bonfire in the main square of the city, burning somewhere in the region of two million books—a "cultural holocaust" based on the principle that, "to destroy the written word is to deprive a culture of its soul, and eventually of its identity."[22] Proclamations followed which banned writing in Arabic and prohibited the ownership of Arabic books. Ximénez de Cisneros was so successful that, by 1609, only a tiny number of Arabic manuscripts existed in Spain. The Catholic victory was complete, "only the empty palaces and converted mosques remained as mute witnesses to the tragedy that had befallen the once flourishing Islamic civilization of Al Andalus."[23]

Luckily for civilization as a whole, many of the most important scientific books were already safely translated into Latin and, as we shall see, making their way across the Mediterranean and Europe to the printing presses, from where their ideas would go on to transform and inform the foundations of the modern world. For many of these books, their transformation from Arabic to Latin took place in a picturesque town, on a rocky hilltop, 300 kilometres north of Córdoba—Toledo, our next destination.

FIVE

Toledo

In the Arab schools of Cordova and Toledo, were gathered, and carefully preserved for us, the dying embers of Greek learning.

—Ahmed ibn Mohammed al-Makkari,
The History of the Mohammedan Dynasties in Spain

SOMETIME IN THE MIDDLE of the twelfth century, a young man arrives at the gates of Toledo. He stands on the edge of the Tagus gorge before crossing the bridge into the city. Far below him, the icy river churns its relentless course through the rock as he gazes across at the city on its granite hilltop. His name is Gerard and he is especially interested in astronomy. Having learned everything he could from his teachers back in Italy, he has travelled hundreds of miles across land and sea from his home in Cremona in search of knowledge. He has been told that, here, in the city of Toledo, in Spain, he will be able to study the discoveries of the Arabs and, if he is really lucky, he will be able to find a copy of the greatest astronomical book ever written: *The Almagest*. He is tired and dusty from many days on the road, but at last his long journey is at an end; he has arrived. Looking across at the dark tangle of narrow streets, he shivers in anticipation of the treasures that await him. As he stands on the threshold of a new chapter in his life, Christian Europe is poised on the brink of a new chapter in its intellectual development. Gerard's work in Toledo will make the city the most important centre for the transmission of scientific knowledge between the Muslim and the Christian worlds; he will spend the rest of his life here, translating books from Arabic into Latin. Copies of these books will travel all over Europe, passed from hand to hand, packed in chests, squeezed into saddlebags, they will jolt along roads from monastery to cathedral school, from university lecture hall to scholar's study. From Montpellier to Marseilles, from Paris to Bologna, Chartres,

Oxford, Pisa and beyond, these books will form the framework of scientific learning for centuries to come. More than any other individual, Gerard of Cremona will be responsible for bringing the great ideas of ancient Greece and medieval Islam to Western Europe.

Gerard's journey to Toledo shows that the fame of Arabic learning had already spread far into Western Europe. The city's position on the frontier between the Muslim and Christian worlds made Toledo, like Palermo in Sicily and Antioch in Syria, a portal through which knowledge flowed. The second half of the eleventh century was a crucial period of revival for Western Europe. The Normans took Sicily from its Muslim rulers and, in 1095, Pope Urban II preached the First Crusade, sending Christians from all over Western Europe to support the Byzantines in their war with the Turks, and then on to take the Holy Land from the Islamic Empire. In the summer of 1099, the crusaders entered Jerusalem and triumphantly returned the Holy City to Christianity. This was the first in a series of crusades against Muslim forces in the East; there were two more in the twelfth century and several in the thirteenth, as the two great religions struggled to expand and hold their spheres of influence, which were centred on Jerusalem. Crusader states were founded and fought over in Antioch, Tripoli and Edessa; some cultural exchange took place here, but the lack of political stability and regular outbreaks of violence meant that the transmission of intellectual ideas was limited, overshadowed by what was happening in Sicily and, even more so, in Toledo.

The Christian reconquest of northern Spain began in earnest in the second half of the eleventh century; Toledo fell in 1085. Within a few decades, scholarly networks were humming with rumours about the wonders that could be found there. These rumours enticed men from the furthest corners of Europe: from England, from Germany,

France, Hungary and the Dalmatian coast. Gerard himself might well have been, directly or indirectly, following in the footsteps of his fellow countryman, Plato of Tivoli, who was translating scientific books, including one by Ptolemy, in Barcelona in the 1120s and '30s. Whether or not Gerard ever actually met Plato, it is certainly possible that he studied his books and that they inspired him to search for more. But there is absolutely no doubt that Gerard travelled to Toledo "because of his love for *The Almagest*."[1]

"Although from his very cradle he had been educated in the lap of philosophy,"[2] Gerard must have begun his schooling in Cremona, probably at home with a tutor, followed by a few years at a local monastery school. From there, he might well have gone to nearby Bologna, already a centre for the study of law at its emergent university—the first in Europe. His education would have included some basic mathematics and astronomy as part of the traditional quadrivium, but while the major textbooks and encyclopaedias he studied mentioned Ptolemy as an authority on astronomy and Euclid on geometry, they did not provide any detail. At least one of his teachers must have been interested in the scientific aspects of the quadrivium, and ignited a passion in young Gerard, who then set off in search of deeper knowledge. Where could he have gone next? The most obvious place in the vicinity was the monastery of Bobbio,[*] just eighty kilometres to the south-west of Cremona, whose library housed one of the only collections of astronomical manuscripts in Western Europe at that time.

In the ninth century, Bobbio had been home to a monk from Ireland, called Dungal, who, in a letter to Charlemagne, was—remarkably—able to explain the science behind the two great solar eclipses of 810. And it was Dungal who amassed a considerable col-

---

[*] The setting for Umberto Eco's classic novel, *The Name of the Rose*, a murder mystery which explores the intellectual world of a fourteenth-century monastery.

lection of manuscripts for the monastery library. By Gerard's time, the collection had expanded to include several works on astronomy by the brilliant eleventh-century monk Hermannus Contractus, who also wrote a treatise on the astrolabe and its uses. A tenth-century catalogue from Bobbio lists a summary of *The Almagest* by Boethius, "the man who made the astronomy of Ptolemy available to the Italians."[3] Another important figure involved with Bobbio was Gerbert d'Aurillac (*c.*945–1003), who went on to become Pope Sylvester II. As a young man, Gerbert's interest in mathematics took him to northern Spain, where he studied under Atto, Bishop of Vich. Whether or not he was exposed to Arabic learning at this time is a matter of contention, but several texts have been attributed to him, on arithmetic, geometry and computing using a special abacus which featured the Hindu-Arabic numerals; he was also interested in astronomy and built armillary spheres. Gerbert was Abbot of Bobbio for a short period, and his letters reveal the role he played in sourcing texts for the library. On 22 June 983, he wrote to Archbishop Adalbero of Rheims, telling him that, "we have since discovered, namely eight volumes: Boethius *De astrologia*, also some beautiful figures of geometry."[4] Five years later, and now in Rheims, he wrote to a friend at Bobbio, "you know with what zeal I am everywhere collecting copies of books," before going on to list books he had had copied.[5]

There were two main routes Gerard could have taken to Spain: south, via Sicily, or north, around the coast of France. Bobbio happens to be on the road from Cremona to Genoa, the closest port and therefore the most likely point of departure for both. The monastery library would have certainly had a reputation for its collection, and for astronomy in particular, so it is not unlikely that Gerard would have paid it a visit at some point before he left for Spain, if not on his way there. It is also clear that it contained various books which could have provided Gerard with the impulse to seek out a full version of *The Almagest*.

Gerard was born in 1114, and he probably spent at least the first twenty years of his life in northern Italy, before setting off on his quest for *The Almagest*. Let's imagine he took the northerly route, setting off from Genoa on one of the many merchant ships that tacked along the coast of northern Italy to southern France, docking in the ports of Antibes, Fréjus and Hyères. In the Middle Ages, travel by boat was the fastest and least uncomfortable form of transport; between April and November, the Mediterranean would have teemed with ships taking passengers and commodities from port to port, always staying close to the coast, where possible. Gerard would have reached Marseilles in just a few days, and if he had disembarked there, he would have found a thriving intellectual scene. Had he decided to stay for a while and study, it is quite possible he would have met an astronomer called Raymond, who was there in 1140, creating a set of astronomical tables for the local area. This is, of course, all conjecture, but well within the realms of possibility. It also provides an answer to the question of why Gerard of Cremona went to look for *The Almagest* in Toledo, and how he knew it would be there. There were various intellectual connections between Toledo and Marseilles—the most important being that Raymond's *Tables* were based on the *Toledan Tables*, created in the previous century by the astronomer al-Zarqali using none other than al-Khwarizmi's *Zij*. If young Gerard spent time in Marseilles, scholars there would have told him about the incredible discoveries of Arabic science, its brilliant scholars and their groundbreaking books. And, if he was not already headed there, they would have certainly pointed him in the direction of Toledo.

Standing on the edge of the Tagus gorge, Gerard would have immediately understood why the founders of Toledo chose the site.

The city sits on top of a steep hill, surrounded on three sides by the sinuous river, which flows through a steep ravine, making it supremely defensible. Attack from across the river would be suicidal—getting down the precipitous cliff face would be hard enough; crossing the fast-flowing water and then climbing up the other side, ready to fight, would be impossible. As the Roman historian Livy put it, "*urbs parva, sed loco munita*"—"a small city, but fortified by its location." Toledo flourished under the Romans as Toletum; the local steel-working industry, famous for its high-quality alloy metal that was extraordinarily hard, supplied the imperial army with swords, and the town grew wealthy. When the Visigoths came to power in Spain, they made Toledo their capital, the nucleus of their political, religious and cultural power, right at the centre of the peninsula. Visigothic learning flourished there in the seventh century as several ecclesiastical writers and at least two libraries made it their home.

This period of supremacy ended abruptly in 712, when the Arabs invaded from the south and established Córdoba as their capital. Toledo, which they called Tulaytulah, remained under Muslim control for several centuries, ruled by local families with varying degrees of autonomy from the Umayyads. As a strategic frontier town, near the border with the Christians of northern Spain, it lay at the very edge of the Arab world. The city declined in the decades that followed, becoming a breeding ground for rebellion and discontent, at the mercy of local warlords, riven by internal strife and subjected to endless sieges. But, after the fall of the Umayyad dynasty in 1031, Toledo became an independent taifa state, and relative stability returned, enabling culture and scholarship to flourish. Vibrancy returned to Toledo's ancient metalworking industry, making the city one of the wealthiest in Spain. Toledan craftsmen were famous for their razor-edged knives, beautiful ornaments and ingenious tools, but most of all for the legendary swords they exported all over the known world, the heart's desire of every ambitious warrior.

In 1029, a young man had been born into an artisan family who lived in a small village on the edge of the city. Al-Zarqali, "the little blue-eyed one," trained, like the other boys in his family, as a scientific instrument maker, and his considerable talents brought him to the attention of Saʿid al-Andalusi, a local judge, teacher and author of the *Book of the Categories of Nations*. In it, Saʿid gives us a lively comparison of the intellectual achievements of various countries, surveying their scholars and contributions to almost every aspect of knowledge. He divides the people of the world into two classes: those who have contributed to science and those who haven't. Unsurprisingly, the chapter on Andalusia is the most interesting and most detailed, but in general it was an influential book and it remains an important source of information on the history of science, complementing al-Nadim's much more comprehensive work, *The Fihrist*.

Under Saʿid's patronage, al-Zarqali made complex instruments for astronomical observations. At the same time, he studied astronomy, and, in 1062, he joined the group of scholars observing the heavens. His technical expertise, combined with his flair for astronomy, meant that he was soon put in charge of the whole project. The most innovative and brilliant producer of astronomical instruments in all of Islam, al-Zarqali's design for a new "universal" astrolabe, called a saphaea, was so revolutionary that it was copied all over Europe, the Middle East, North Africa and even as far away as India. He also created other marvels—people came from far and wide to see his marble water clocks, which told the time with unheard-of accuracy. Al-Zarqali studied in Córdoba, but returned to Toledo, where he wrote several books, including one called *The Canones* (rules) that explained how to use the *Toledan Tables*. Gerard of Cremona translated this work into Latin and it went on to influence European astronomy for centuries. Al-Zarqali also wrote an astronomical treatise in which he made the groundbreaking claim that Mercury's orbit was elliptical, not circular, as was commonly thought. In the

15. An astrolabe made in Toledo in 1029 when the city was still under Muslim rule.

sixteenth century, Johannes Kepler drew on al-Zarqali's visionary work to prove that the orbit of Mars was also elliptical. Toledo fell to Alfonso of Castile in 1085, and al-Zarqali left the city, but his ideas were taken up by Christian scholars and thus shared throughout Europe.

Al-Zarqali moved south, possibly to Granada or another Andalusian city still under Muslim control, as did many of his fellow Arabs. The Mozarabs, who had remained faithful to Christianity throughout four centuries of Islamic rule, worshipping according to rites they had inherited from the Visigoths, stayed behind and watched as the new rulers began to impose the Catholic, Latin ritual, derived from Rome. It must have been a strange time for this long-established people. Like Andalusia's Sephardi Jews, they had created their own society within Muslim Spain, maintaining their religious beliefs, but adopting the language, dress and characteristics of their overlords—a hybrid community emblematic of and dependent on the multicultural nature of the place in which they lived. On the one hand, Toledo's Mozarabs probably felt some relief that Christianity had triumphed and returned to their lands; on the other, there must have been sadness at the loss of Muslim friends and colleagues, and concern about what the future held under the Castilian king. This was entirely justified: over the next four hundred years, Catholic Spain gradually absorbed Mozarabic culture by confiscating their lands and refusing to recognize them as a separate legal community. Some isolated vestiges did survive. In 1502, copies of the Mozarabic liturgy were collected together, and a chapel in Toledo Cathedral was dedicated to their faith—it is still there.

The Mozarabs occupied a unique territory between two cultures: Christians under Muslim rule, who embraced Arab customs, but still spoke their own language and lived by their own laws. It is ironic that they managed to survive so successfully and for so long under a rival faith, but were then persecuted by Catholicism, a different form of their own religion. This says as much about Mozarabic tenacity as it does about medieval Islam's capacity to accommodate alternative religions within its own sphere of influence. It was a similar story with the Jews—persecuted by the Visigoths, they thrived under the Umayyads, then were exiled and murdered by the Catholic Inquisition. But not all the ruling Muslim dynasties were tolerant. The Almohads and the Almoravids, who ruled large parts of the Iberian peninsula in the eleventh and twelfth centuries, persecuted Jews and Mozarabs alike, causing many of them to flee north, to Christian Spain. But, apart from some initial hostility from the Frankish clergy, it was not until much later, in the fifteenth century, that the victimization of the Mozarabs and the Jews began. At first, these two communities continued to flourish in Toledo, especially in the scholarly sphere, where their linguistic skills and knowledge of local libraries were invaluable.

With Toledo once again under Christian rule, the Catholic Church needed to establish its religious dominance. In the tenth century, the Black Monks of the Order of St. Benedict had spread from the Abbey of Cluny, in the Loire, down through France and across the Pyrenean valleys of northern Spain. And it was from this monastic order that Toledo's clergy now came. The Benedictines settled in the streets around the cathedral, and, in the decades following the reconquest, the "Frankish quarter," as it became known, was where newcomers—clerics, scholars, foreigners—gathered to live, work and share their ideas. As a result, a busy line of communication and travel opened up between Toledo and France, and especially the cathedral schools of Paris and Chartres.

This was the cultural scene into which Gerard launched himself in the mid twelfth century. He would have crossed the gorge over the old Roman Alcántara Bridge and then begun the steep climb up the narrow alleyways into the city. It is easy to imagine the Toledo he found himself in, because it has changed very little since. The narrow alleyways are still steep and shaded, the shops still sell a dizzying array of knives and swords, the glittering blades neatly fanned out on velvet cushions, guarded by fearsome suits of armour. Marzipan is still made from almonds grown in the orchards that surround the city—a tradition begun by the Arabs when they introduced sugar palms to the region. But Gerard would not recognize the cathedral in modern Toledo; it was built after his death. The cathedral in which he worked and worshipped was, in fact, a mosque, converted into a church after the Christian reconquest of the city, and on the same site as the current cathedral. Gerard probably arrived with letters of introduction, and would have gone to the Frankish quarter to find somewhere to stay and make enquiries about local scholars. His search for Ptolemy's great masterpiece could now begin in earnest. His first port of call could well have been the cathedral library, but it probably did not contain much of interest to him. He would have had to look further afield, to other libraries in Toledo, many of which had survived from Islamic times. These collections were home to a huge wealth of Graeco-Arabic scientific texts, which European scholars had already started to study and translate into Latin. We can only imagine Gerard's excitement when he finally found himself sitting at a desk with *The Almagest* in front of him, and the hurry he must have been in to learn Arabic so that he could understand and then translate it. Gerard's students wrote a short biography of him and included it in the Preface to the translation he made of Galen's *Tegni*. In it, they explained that "seeing the abundance of books in Arabic on every subject, and regretting the poverty of the Latins in these things, he [Gerard] learned the Arabic language in order to be able to translate."[6]

This was no small task. Arabic is an extremely complex language with a different alphabet and direction of writing, and an intricate system of diacritics, but there were plenty of Mozarabs around to help. A man called Ghalib assisted Gerard with his translation of *The Almagest*—he probably taught him Arabic at the same time. Toledans spoke a local variant of the Iberian Romance languages, the precursors of modern Spanish. Gerard would have doubtless learned this as well, so that he could communicate easily with Ghalib and other local people. Jewish scholars, trilingual in Hebrew, Arabic and the local Romance language, formed another important bridge between the two cultures, helping with translation and providing continuity between the Arab past and the Christian present. Though many Muslim Toledans had moved south when the city fell to Alfonso VI in 1085, some remained, and relations between the two communities were often close. In fact, northern Muslim families often allied with the Christians against the hard-line Islamic Almoravid dynasty, which was making its presence felt in al-Ándalus.

One such family was the Banu Hud, who ruled the city of Zara-goza from 1039 to 1110. Keen scholars in their own right, they built up an impressive library of scientific texts. Yusuf al-Mutamin ibn Hud, who was mentioned in the previous chapter and who ruled Zara-goza from 1081 to 1085, was a notable mathematician—probably the most innovative in the whole of Muslim Spain. He wrote a com-prehensive book on geometry, called *The Perfection*, that was based on texts in his library, including Euclid's *Elements* and *Data*, Apol-lonius' *Conics*, and Archimedes' *On the Sphere and the Circle*. In 1110, the Banu Hud family lost Zaragoza to the Almoravids. As a result, they allied themselves with Alfonso I, the Christian King of Ara-gon, and moved to Rueda de Jalón, near Tarazona, in the Ebro Val-ley. They were on good terms with their Christian overlords (the last Banu Hud ruler, Saif al-Dawla, was a guest at Alfonso's coronation), even after the Christians took Zaragoza for themselves from the Almoravids, in 1118. Michael, Bishop of Tarazona from 1119 to 1151, an avid collector of astronomical texts, chose manuscripts from the Banu Hud library for Hugo Sanctallensis to translate for him. Rueda de Jalón was not far from Tarazona, and Hugo describes finding the manuscript of a commentary on al-Khwarizmi's *Zij*[7] there, "among the more secret depths of the library."[8] What else he found is a mat-ter of speculation, but given the Banu Huds' personal interest in sci-ence, their library was almost certainly a major source of books for the scholars and translators of the twelfth century. In 1141, the fam-ily were forced to exchange their lands for a house in the cathedral quarter of Toledo. Presumably, they brought their books with them when they moved, and, if so, put them in easy reach of the translators who later made the same part of the city their home. Indeed, Gerard of Cremona translated several texts on geometry that al-Mutamin had drawn on in *The Perfection*.

The Banu Hud library is important because, thanks to Yusuf al-Mutamin ibn Hud, we can be certain about some of the books it contained, but there were already many other libraries in Toledo about which we know much less. The city was an important centre of learning during the tenth and eleventh centuries, and when the Christians took over in 1085, the transfer of power was peaceful. As a result, even though the majority of the Muslim elite emigrated south, their culture was preserved, libraries were protected and the various communities of Jewish, Arab, Mozarabic and Christian scholars were able to work together. This was especially important for the programme of translation from Arabic to Latin (often via Hebrew or Romance) that followed. In the early Middle Ages, Spain was a multilingual society. Under Muslim rule, Arabic was the language of education and government, but Romance was spoken on the streets and in the fields, intermingled with various Berber dialects. Latin was the language of the Mozarabic Church, and of course Hebrew was ever-present in the large Jewish communities. When Toledo was reconquered by the Christians, Latin, the language of the Catholic Church, took on an increasingly important role, but the Mozarabs continued to use Arabic until well into the fourteenth century.

The European scholars who came to Toledo soon after the reconquest were staggered by the wealth of knowledge they found there. In the medieval period, Arabic book culture positively dwarfed that of Western Europe; the twelfth-century scholar Bernard of Chartres was proud of the twenty-four books he owned, but, in 1258, the city of Baghdad boasted thirty-six public libraries and over a hundred book merchants. The largest medieval library in Christian Europe,

Toletū ciuitas bispanie

16. A fifteenth-century engraving of Toledo.

at the Abbey of Cluny, contained a few hundred books, while the royal library of Córdoba had 400,000. Even if we allow for exaggeration and the fact that the Arabs still mainly used scrolls, which could not contain as much text (several would be needed for one copy of a codex), and that paper was not produced in Western Europe until the fourteenth century, so it had to be imported, making books more expensive, the comparison is still shocking. Arab textual culture was not only much larger, it was also infinitely richer. The scale of Arab accomplishment in literature, history, geography, philosophy and, of course, science left Latin scholars dazzled, giddy with awe. There was a lot of catching up to do.

Historians have argued long and hard over how the translation movement in Toledo functioned. Was there a school of translators working together? If so, where were they based? Who paid for the translations? Who chose what was translated, and how did they choose? The huge quantity of scientific material on offer meant that selection was a difficult yet unavoidable process. As usual, lack of evidence holds us back from making definite claims. According to his followers, Gerard of Cremona translated "books of many subjects—whatever he esteemed as the most choice."[9] This implies that he was responsible for choosing the texts he worked on, which is totally plausible, given his expertise. A list of the books he translated during his lifetime numbers seventy-one—and others have been identified since. They are divided into groups: dialectic (logic)—three; geometry—seventeen; astronomy—twelve; philosophy—eleven; medicine—twenty-four; and miscellaneous—four. The subject headings give us a clue as to how Gerard might have been organizing his translation programme—they are loosely based on the liberal arts that underpinned the ancient Greek curriculum adopted by Arab scholars as the basis for their education system. Collections of texts were compiled to use as teaching materials for students, and Gerard seems to have deliberately sought out collections in mathematics, astronomy and medicine so he could make them available to students in the West. One such collection was called the *Middle Collection* or *Little Astronomy* because it was designed to be studied between *The Elements* and *The Almagest*.

Once students had mastered the basics, they could move on and study their subject in greater depth. For this, they needed the full versions of *The Elements* and *The Almagest*, and other related works. By AD 500, scientific texts in the Latin world had been reduced to brief extracts in handbooks and encyclopaedias, "condensed into small packages . . . for the long journey through the Dark Ages," as the great twentieth-century historian Charles Homer Haskins put it.[10] Gerard and his peers seemed to be deliberately trying to unwrap those small packages in order to broaden and deepen scholarship and thus revitalize education. The encyclopaedias and the digests were no longer enough. They had to go back to the great textbooks of antiquity and translate them in full. It also meant translating the works of Arabic scholars that explained and built upon those ancient texts, many of whom we have already met on our journey: al-Zahrawi, al-Razi, al-Kindi, the Banu Musa and al-Khwarizmi, among others.

One of the dilemmas facing translators like Gerard was whether to focus on the full versions of ancient Greek texts, or prioritize the corrected, improved Arabic versions, brilliantly synthesized with ideas from Persia, India and Egypt. As ever, personal choice played a vital role in which works were handed down to posterity, and which were not. Gerard went for a combination of the two, and loosely based his selection of texts on *The Classification of the Sciences* by the great philosopher al-Farabi (872–950), who had spent most of his life in Baghdad, where he was known affectionately as the Second Master (to Aristotle's First).

Where would Gerard have found all these manuscripts? We have already looked into the Banu Hud library, but what about the others in Toledo? It is difficult to assess private collections, but the thriving community of scholars in the city during the last century of Arab rule (985–1085) would have doubtless owned copies of the most important texts. Sa'id al-Andalusi talks about the scientist Abu Uthman Sa'id ibn Muhammad ibn Baghunish, who was from Toledo, but had studied in Córdoba before returning to his home city to become an administrator at the court of the al-Nun rulers:

> a cleanly dressed and pious man, who had in his possession great books on the various branches of philosophy and other fields of knowledge. I came to realize by talking with him that he had studied geometry and logic and that he had precise knowledge of both fields, but he neglected this area to give special attention to the books of Galen, of which he had a private collection that he

had critically corrected, thus becoming an authority on the works of Galen.[11]

Here is a rare, first-hand piece of evidence of Galenic books in a private collection in Toledo. It would not be unreasonable to suggest that *The Elements* and *The Almagest* might well also have been on ibn Baghunish's shelves, and that at least some of his books—or copies of them—could have still been in the city in the mid twelfth century when Gerard was on the lookout for texts to translate. There were certainly still Arabic collections in the thirteenth century. The scholar Mark of Toledo (*fl.*1193–1216) claimed that he "studiously sought another book to translate in the libraries of the Arabs."[12] Mark was fluent in Arabic, and his particular area of interest was Galen, and, having studied medicine abroad, he returned to his home town to find and translate Galenic treatises that were not yet known in the West,[13] contributing to a widespread revival of Galen's work in the thirteenth century. Mark filled in the gaps left by Gerard, who had translated nine works by Galen. Of the twenty-four medical texts Gerard translated, by far the most significant was Avicenna's *Canon*, itself a synthesis of Galenic medicine, and the most popular medical textbook in the Middle Ages. Avicenna (Ibn Sina, 980–1037) was one of the greatest thinkers of the medieval Arabic world. Described as "a brilliant summary and logical restructuring of Galenic medicine,"[14] his *Canon* was a manageable five-volume affair, much more practical and affordable than the sprawling Galenic corpus, and it became the primary channel for the transmission of Galen's ideas. Gerard also translated several of al-Razi's books, which were transmitted to Europe as a collection and later printed, as was al-Zahrawi's treatise on surgery and instruments, complete with beautifully copied illustrations and diagrams.

So far, we have looked at where translators might have found books in Christian Spain, but there is evidence that scholars were also looking in the south, where the Muslims were still in power. An early-twelfth-century source describes an Andalusian market supervisor enforcing an order that proclaimed, "Men should not sell scientific works to the Jews or the Christians,"[15] because they were apparently translating them and then attributing them to their own scholars. This episode reveals that Christian interest in Arabic science had become so pronounced and acquisitive that the Muslim authorities had begun to fear that their cultural inheritance would slip away over the border to the north, a fear that turned out to be prescient. We do not know how widespread or successful this policy of restricting sales was, but it must have made getting hold of texts more difficult.

It is hard to say, though, how much of an effect this had on Gerard's project. He was certainly able to acquire texts, no matter the prohibitions of Andalusian market supervisors. The scale of what he achieved tells us something about his personality. Gerard of Cremona must have been an extremely bold, determined man. Anyone who sets off into the unknown in search of a single book, who is prepared to travel thousands of miles, to learn at least one entirely new language and spend the rest of his life in a strange country, relentlessly seeking to broaden his knowledge, must have had a fairly specific set of personal characteristics. Intrepid, driven, diligent, brilliant—these qualities contributed to the fact that, "more of Arabic science in general passed into Western Europe at the hands of Gerard of Cremona than in any other way."[16] If we believe one account of a lecture Gerard gave in Toledo, we can add arrogant and pompous to the list. He was clearly aware of the significance of the job he had undertaken, and the importance of his position as a major conduit for the transfer of knowledge from the Arab world to the European. As the writers of his eulogy put it, "to the end of his life, he continued to transmit to the Latin world (as if to his own

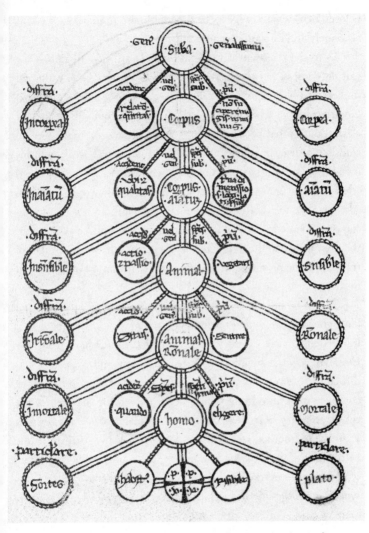

17. Diagram from a manuscript containing Gerard of Cremona's translation of al-Zarqali's *Canones*.

beloved heir) whatever books he thought finest, in many subjects, as accurately and as plainly as he could."[17]

As to the question of who paid for Gerard's translations—we can only guess, but the fact that he was a canon of the cathedral makes Archbishop John of Toledo (1152–1166) a likely candidate. Gerard could have been on the cathedral payroll, but with such light clerical duties that he was free to devote the majority of his time to translation. It is, of course, also possible that he was wealthy in his own right, and able to fund himself. According to his pupils, Gerard "fled fawning praises and the empty pomp of this world," and was "an enemy to the desires of the flesh."[18] Frivolous he was not, rather someone who was entirely focused on his work and had no use for the finer things in life. It is certainly clear from the sheer number of translations he produced that he must have spent most of his time at his desk.

Gerard went to Toledo to find *The Almagest*, but he would not have got very far with Ptolemy's great work before realizing that he needed to get to grips with Euclid first. As we have seen, *The Elements* was the stepping stone to astronomy and, "As such it is difficult to over-emphasize the importance of the translations of the complete work into Latin which were made in the first half of the twelfth century."[19] In 1100, all that was available of *The Elements* in Latin were fragments of Books 1–4, translated by Boethius in the fifth century, with hardly any proofs or diagrams. By 1175, there were six new versions of the complete text; Christian Europe had woken up to the importance of Euclid's theories and Latin scholars were working hard to comprehend and transmit them.

Gerard's Latin edition of *The Elements* was the second of three translations from Arabic in the twelfth century. The first was by Adelard of Bath, and it formed the basis of the third version, produced by Herman of Carinthia and Robert of Ketton. Gerard seems mainly to have used the Ishaq/Thabit version, with certain parts of

Hajjaj's translation. This might mean that he had two Arabic versions in front of him, but equally he might have been copying a text that was already an amalgam of the two; as we saw in the previous chapter, combinations of the two texts had existed in a myriad different versions since immediately after their original creation, in the ninth century. Because it was primarily based on the Ishaq/Thabit version, Gerard's text is closest to the original Greek and even features some Greek words. He also included all Euclid's proofs in full—an important departure from other versions and one that made it easier to understand what Euclid had written. Gerard's usual method was to make a literal translation, with each word translated individually, rather than trying to convey the general sense of the text—a method common to the circle of translators in mid-twelfth-century Toledo. Interestingly, Gerard's *Elements* was slightly more influential than Herman's, but both were completely overshadowed by Adelard's version, which was far more widely disseminated and has survived in many more manuscripts, although the reasons for this are unclear. It formed the basis of the revised text produced by the great thirteenth-century Italian mathematician Campanus of Novara, which was the first version to be printed, in Venice, in 1482.

If Gerard followed the accepted order for the study of mathematics and astronomy, he would have translated *The Elements* first, before he embarked on the *Middle Collection* (the texts that were supposed to be read in between) and then finally started on *The Almagest*. It is possible that he was unable to disseminate his work very effectively early on in his career, and that, as time went by, his contacts improved, hence his version of *The Almagest* was much more influential than his *Elements*. Regardless of the extent of their individual influence, all the twelfth-century editions of *The Elements* contributed to an increasingly lively debate about Euclidean mathematics that continued well into the thirteenth and fourteenth centuries. The theories themselves were transmitted in many different ways: full translations, partial translations, compilations, Latin versions of Arabic commentaries, new Latin commentaries and original works by Latin scholars. Seven manuscript copies of Gerard's *Elements* survive today: one each in Oxford, Boulogne, Bruges and Paris, and the remaining three in the Vatican Library. None of them mention Gerard by name, and almost all of them were made during the fourteenth century. Only four are complete versions of Gerard's translation; the others contain a mixture of texts from the various editions made in the twelfth century, which shows that whoever copied them had access to several manuscripts.

As we know from our time in Baghdad, mathematics had broadened out dramatically to include much more than just Euclidean geometry, and the other mathematical books Gerard translated reflect this. Al-Khwarizmi had opened up the study of arithmetic using Hindu-Arabic numerals and the decimal system in *The Book of Addition and Subtraction According to the Hindu Calculation*, and had defined algebra as a distinct subject in the *al-Jabr*. Both of these texts were reproduced in Latin in various forms from the twelfth century onwards. The *al-Jabr* was translated by Robert of Chester (often wrongly confused with Robert of Ketton) in Segovia in 1145, and also by Gerard, in Toledo, while the Latin versions of Khwarizmi's book on arithmetic helped to spread the decimal system to Europe.

In the archives of Toledo Cathedral, there are two documents which refer to a Master Gerard (*magister*). Both are signed by Dominicus Gundissalinus, a local cleric who worked alongside Gerard somewhere in the precincts of the cathedral. The only clue as to where they worked comes from a scholar in the following century, who apparently produced a translation, "in the chapel of Saint Trinity"—a monastery next to the cathedral.[20] Like the House of Wisdom in Baghdad, there is no surviving evidence for a physical place set aside for translators, but common sense suggests that a translation programme of this scale must have had some kind of central location, a permanent space where they could keep their books and the other scribal equipment they used: pens, ink, knives, paints, paper or parchment leaves and glue, with desks to work at in peace and quiet. There is certainly evidence that they sometimes worked on translations in pairs. We know that Ghalib the Mozarab helped Gerard with his Arabic, and another translator, John of Seville and Limia,[21] described his work thus: "The book ... was translated from Arabic, myself speaking the vernacular word by word, and the archdeacon Dominic [Gundissalinus] converting each into Latin."[22]

Whether or not they actually sat side by side in the same room, Dominicus had different intellectual interests, or at least operated in different scholarly areas to Gerard. His translations focused on Arabic philosophy, especially the work of Avicenna, and he collaborated with local Jewish scholars, using Hebrew commentaries and ideas. It is possible that Gerard and Dominicus consciously divided up these areas of knowledge between them, to streamline the translation process. John of Seville mainly concentrated on astrology—a subject Gerard does not seem to have translated at all.[23] Does this division of labour constitute an organized programme of translation? And, if so, who was in charge of it? Gerard did not dedicate or sign his translations, so leaves us with no clues. But Herman of Carinthia and Robert of Ketton are more helpful. Their prefaces are littered with dedications and evidence of their motives. Herman's main patron was the great teacher and philosopher Thierry of Chartres, with whom he had studied before travelling to Spain. Thierry was interested in astronomy and mathematics—he used Pythagorean theory to explain the Trinity and compiled *The Heptateuchon*, or *The Library of the Seven Liberal Arts*, in which he included sections of *The Elements* from the translation by Herman and his colleague Robert of Ketton, which they had sent to Thierry for *The Heptateuchon*. They sent other works, too, including Ptolemy's *Pla-*

nisphaerium, a mathematical exploration of how to map stars onto the celestial plane.

Thanks, in part, to the books Herman sent him from Spain, Thierry of Chartres was at the forefront of the European discovery of Graeco-Arabic science. Herman had studied under Thierry as a young man, travelling to Chartres from his native Istria to do so, and they stayed in touch. Herman and his fellow student, Robert of Ketton, spent many years travelling in search of texts, which they translated and sent to France with enthusiastic letters praising the marvels of Arabic scholarship.

When the Abbot of Cluny, Peter the Venerable, was touring Spanish monasteries in 1141, he came upon the two young scholars, "on the banks of the Ebro," and persuaded them to translate the Qur'an for him.[24] Peter used this and other Muslim texts to posit the notion that Islam was a Christian heresy. More importantly, though, he was the first Christian scholar to engage with and learn about Muslim culture. This was a landmark moment in the dialogue between the two religions. Peter's interest in and respect for Islam was evident. He described Muslim scholars as "clever and learned men whose libraries are full of books dealing with the liberal arts and the study of nature, and that Christians have gone in quest of these."[25] This interested, open attitude was characteristic of European scholars in the twelfth century, but it did not last. By the time of the Renaissance, the Muslim scholars who had made such an enormous contribution were being obscured by an obsessive reverence for ancient Greek sources.

Having translated the Qur'an, Robert was keen to get back to astronomy, promising Peter in 1143, "a celestial gift which embraces within itself the whole of science. This reveals according to number, proportion and measure, all the celestial circles and their quantities, orders and conditions, and, finally, all the various movements of the stars . . ."[26] This is a reference to *The Almagest*, which he and

Herman had been preparing to study by reading *The Elements* and other mathematical texts. It is possible that they produced their own version of Ptolemy's text, but there is no evidence of this. In contrast, Gerard's version of *The Almagest* was the first to be widely disseminated in Europe and was the most influential of all the early Latin versions of Ptolemy's work: fifty-two manuscripts survive today, in libraries as diverse as St. Petersburg and Manchester. By far the largest group—thirteen copies—are in the Bibliothèque Nationale in Paris, which could indicate that Gerard's Frankish colleagues in Toledo sent copies back to France. The oldest of his manuscripts to survive is in the Vatican Library, with another four, later copies, which suggests that Gerard's translations travelled to Italy, where they were reproduced in considerable numbers. By the time it appeared in Latin, *The Almagest* was already quite well known, at least among the tiny, elite group of scholars interested in mathematical astronomy. It had been referenced and quoted in several other works and was a major source for Herman of Carinthia's *De essentiis*, which had introduced *The Almagest* to Western Europe seventeen years before the first complete translation was produced.

A major issue faced by all translators of scientific texts, those in Baghdad as much as those in Toledo, was the need to create a vocabulary to explain new technical ideas. This was especially true for astronomy and mathematics, where the innovations of the Arabs combined with the lack of experience of Latin culture meant that most of the concepts and methodologies being translated were entirely original, and there were—as yet—no words to express them. Jewish scholars, already used to conveying these ideas from Arabic into Hebrew, made a significant contribution and helped Latin scholars to create new terminology.

As we have seen, some scholars based in Spain sent copies of their translations to monasteries and cathedral schools in France, where they were studied and re-copied, and thus disseminated through the extensive Benedictine network. But there was not much demand for them in Spain itself, where universities and schools developed more slowly—the first Spanish university was not founded until 1208. There was an early manuscript copy of Gerard's *Almagest* in Toledo Cathedral Library, now in the National Library of Spain, in Madrid,[27] but it is the only one in the whole of Spain. It is written on parchment by a single scribe in neat handwriting, and was produced, probably in the cathedral itself, at some point in the thirteenth century, so there must have been a manuscript available to copy it from. But the library did not have a significant collection of scientific works until the late thirteenth century, which raises the obvious question: what happened to Gerard of Cremona's books? As far as we know, he did not have a patron waiting for them in Italy. He died in 1187, probably in Toledo. There is evidence that he might have bequeathed some of his books to a convent in Cremona, and there is one instance of a scholar, teaching in Cremona in 1198, who appears to have had a copy of Gerard's translation of Galen's *Microtegni*, so there was clearly a channel of transmission of some kind or other.

The most definite piece of evidence we have for books being taken from Toledo to Northern Europe is Daniel of Morley's claim that he brought a "precious multitude of books" back with him to England in the late twelfth century.[28] Daniel was a confident member of a new generation of scholars, who formed small yet powerful networks spanning the continent, and travelled long distances in search of books and ideas. As he himself explained:

> When, some time ago, I went away to study, I stopped a while in
> Paris. There, I saw asses rather than men occupying the chairs
> and pretending to be very important. They had desks in front
> of them heaving under the weight of two or three immovable

tomes, painting Roman Law in golden letters. With leaden styluses in their hands they inserted asterisks and obeluses here and there with a grave and reverent air. But because they did not know anything, they were no better than marble statues: by their silence alone they wished to seem wise, and as soon as they tried to say anything, I found them completely unable to express a word. When I discovered things were like this, I did not want to get infected by similar petrification . . . But when I heard that the doctrine of the Arabs, which is devoted entirely to the quadrivium, was all the fashion in Toledo in those days, I hurried there as quickly as I could . . .[29]

While he was in Toledo, Daniel claimed to have heard Gerard lecturing on astrology, but in spite of this, he did not use Gerard's translations in his own work, preferring those by Adelard of Bath, Herman of Carinthia and John of Seville. The very fact that he had access to different versions of texts and could pick and choose shows how fast the Latin intellectual landscape was changing. A hundred years earlier, at the beginning of the twelfth century, *The Elements*, *The Almagest* and most of the Galenic corpus had not been available in Latin; now, there were several editions, with new revisions and commentaries appearing all the time.

Daniel returned to England with his chests of new books and began making his way to Northampton. On the journey, "I met my lord and spiritual master, John, Bishop of Norwich, who showed me great honour and respect."[30] The most likely location for this meeting was Oxford, a major crossing point on the River Thames and home to the fledgling university. John came from Oxford and definitely returned to visit during his time as Bishop of Norwich. In addition to this, copies of several of the books Daniel brought back from Spain with him appeared in Oxford libraries soon afterwards, suggesting that Daniel let scholars make copies while he was there. It's certainly hard to imagine a man like him resisting the temptation to show off the treasures he had collected.

Cosmopolitan, adventurous, pioneering, Daniel was part of an elite group of highly educated men whose thirst for knowledge drove them to travel widely in pursuit of different horizons, be they geographical or intellectual. Rejecting traditional dependence on written authority and the tendency to ascribe the mysteries of the natural world to divine providence, they adopted new approaches to learning based on rational enquiry and other ideas they discovered in recently translated works of Arabic and classical origin. The works of Aristotle played an important role, and as they were translated into Latin, the full breadth of his cosmographical framework was revealed, something Charles Burnett has described as, "a momentous episode in the history of Western European science."[31] Plato's *Timaeus* was another source of inspiration, prompting scholars to examine their environment in a more rational, analytical way and present their discoveries logically, within an organized scheme, based on the demonstrative method so elegantly espoused by Euclid in *The Elements*. Herman of Carinthia and Daniel of Morley both wrote original works that deployed Aristotelian logic, the demonstrative method and observation of the natural world, as did several of the other scholars we will meet later.

There are many similarities between what was happening in Toledo in the twelfth century and what had happened in Baghdad in the ninth. Knowledge was collected, categorized, translated and organized into distinct disciplines, each with its own specific style, ideas and vocabulary. This flowering of intellectual endeavour was propelled by the same underlying developments in society that had characterized Arabic culture three centuries earlier: diverse territories united under a common religion; an increase in population, agricultural production and trade; and the growth of urban centres, resulting in a demand for infrastructure and regulation, and there-fore numeracy and literacy. In Europe, this process was marked by the growth of secular teaching, visible in the development of univer-sities as the major centres of learning in the thirteenth century. As the old cathedral schools declined, teachers in Oxford, Bologna and Paris competed with one another for students, encouraging intellec-tual rigour, new ideas and the birth of the modern system of higher education. As new disciplines emerged, existing ones grew so that scholars had to be increasingly specialized. It was no longer possible for a student to master the full breadth of knowledge. Toledo played a prominent role in this. The city was a bridge between Graeco-Arabic culture and Latin Europe, a place where scientific knowl-edge was not only held in safety, but translated and transmitted to the scholars of the future. And, in the thirteenth century, Alfonso X (1221–1284)—"the wise," as he was known—ensured that the city continued to be a beacon of learning and cross-cultural coopera-tion. He established a school of Jewish, Christian and Muslim schol-ars to translate important texts into the local vernacular Romance language, and enthusiastically encouraged a programme of astro-nomical study and observation. The result was the *Alfonsine Tables*, based on the earlier *Toledan* ones, which would be used throughout Europe for the next three hundred years.

Toledo was the main site of translation from Arabic to Latin, but it was not the only one. By the end of the twelfth century, the politi-cal and intellectual worlds had changed. Christian Europe was on the rise; Islam was pushed further and further south, down through Spain, back to North Africa, out of Sicily and away from Jerusalem. During the twelfth century, crusading armies fanned out across the Eastern Mediterranean, conquering territories and bringing with them a new sense of confidence and possibility. Merchants followed in their wake: opportunistic men from the Italian city states, their eyes fixed on fortune, settled in Middle Eastern cities and founded trading communities. They dealt, bartered and bargained the treasures of the East—spices, silks, jewels, carpets, antiquities and manuscripts—sailing their ships full of beauty, wonder and wisdom home up the wide shipping lanes, transforming European tastes, style and scholarship for ever.

SIX

Salerno

Tum medicinali tantum florebat in arte,
Posset ut hic nullus languor habere locum.
[Salerno] then flourished to such an extent in the
    art of medicine
that no illness was able to settle here.

—Alfano I, Archbishop of Salerno

IN THE MIDDLE of the eleventh century, a North African merchant called Constantine arrived in the busy Italian port of Salerno with a consignment of goods. During his stay in the city, he fell ill, so a local doctor was summoned to his bedside. We don't know what treatment this doctor prescribed or whether he contributed to Constantine's recovery, but we do know that Constantine was shocked by the doctor's incompetence—the man hadn't even asked for a urine sample on which to base his diagnosis. Many people would have shrugged off the experience, thanked God they had survived, and hurried back to their more civilized homeland. Fortunately for the development of European medicine, Constantine was an unusual character. He remained in Salerno and began to question local people about what kind of medical books were available there, only to be horrified when the answers indicated how little the Italians knew about medicine.

If this story had happened in any other city in Europe at that time, it would have been unremarkable—understandable, even. Medical knowledge was rudimentary in Western Europe in the second half of the first millennium. Doctors were few and far between. There were some medical texts in Carolingian monasteries and in the libraries of the imperial court, but not many people had read and understood them. When they were sick or injured, the overwhelming majority relied on a combination of prayer, blind hope and (often) unsavoury amulets bought from local healers.* The sublime irony of Constantine's story is that he was lucky enough to fall ill in Salerno—then the most advanced centre of medical learning in the whole of Europe. The physician whose knowledge he scorned was, in fact, at the cutting edge of Western clinical science, and had been trained by the best physicians, using the most up-to-date texts available. Salerno's reputation was such that it was known, in Europe, at least, as the *Civitas Hippocratica*. But Constantine's disdain for the treatment on offer eloquently demonstrates the chasm that existed between European scholars and their Muslim counterparts. The Arabic-speaking world had benefited from centuries of medical scholarship and research. They had hospitals, practical guides, diagnostic tools and specialized equipment to treat the sick and the injured. When Constantine discovered that the medical treatment he had received was better than anything else on offer in Europe, he resolved to help.

Constantine returned to his native Carthage, where he began to study and then practise medicine. A few years later, he again sailed to Salerno, bringing with him a cache of books that would transform the study and practice of medicine in Europe, and enhance Salerno's position as the major centre of medical teaching. The Schola Medica Salernitana, founded in the ninth century, is often heralded as the first medical school in Europe. This is misleading. First of all, it wasn't an organized institution in the modern sense, more an amorphous collection of medical teachers around whom students gathered. Just as with Baghdad and Toledo, there is no evidence for a centralized location for medical study in the city. Secondly, there were plenty of medical "schools" in Europe in the ancient world, so any claim for Salerno's status as the first needs to be qualified— "the first medical school in post-antique Europe" doesn't have quite

---

* The Carolingian Empire (800–888) encompassed much of modern-day Germany, France and northern Italy, and its main cities were Frankfurt and Aachen.

the same ring to it. Nevertheless, from about 850 onwards, Salerno was at the forefront of medical study for several centuries, and the school was extremely influential in the spread of medical knowledge to other intellectual centres, and, eventually, in medicine becoming an officially recognized part of the university curriculum.

Historians have puzzled over why and how Salerno became a centre of medicine, where the knowledge came from and how the tradition developed there. Inevitably, there are various possible sources and answers. The most obvious is its geographical position by the sea, with clean air, a temperate climate, beautiful scenery, local hot springs and plenty of fresh local produce—bathing and diet were an integral part of medical therapy. This stretch of the Tyrrhenian coast has been popular with people seeking rest and relaxation for millennia. In around 600 BC, the Greeks had settled a few kilometres further south, and the monumental temples they built at Paestum still stand. By the first century AD, the coast was home to luxurious villas and vineyards, pleasure palaces where senators and emperors came to enjoy the summer and escape the heat of Rome.

Evidence for specific medical study, rather than a general tradition of therapy and well-being, is harder to pin down. Scholars looking for a place near Salerno, where the interest in medicine could have come from, have suggested Velia. A Greek city, eighty kilometres to the south, it had become famous as a centre of learning under Roman rule, in the first century AD, particularly for the study of philosophy and medicine. Archaeologists have found statues of Asclepius and Hygieia, in addition to inscriptions, coins and surgical instruments—all of which support Velia's claim to being a centre of medical excellence. The problem is getting from Velia in the second century (when the city began to decline because of continual outbreaks of malaria) to Salerno in the eighth or ninth century. Of course, it is possible that doctors fleeing Velia went to Salerno, taking their books with them, and started a tradition of medical study that survived the chaos of the intervening centuries, with the result that medicine was alive and well in AD 700 (when the first evidence of medical practice in Salerno appears), but this is impossible to prove.

The best way to approach the early history of the medical school at Salerno is to look at the texts the students were taught from and to trace their origins. These included an incomplete copy of Dioscorides' *De materia medica*, Hippocrates' *Aphorisms* and Galen's *Letter to Glaucon*, all in Latin translations. The last two were part of the Alexandrian curriculum, which had been developed for the education of young doctors in the second and third centuries, and which—by 500—had spread to Constantinople and Athens as well. By 531, they had also found their way, along with various works by Aristotle, to Syria, where they were translated into Syriac and studied by local scholars.* In the 630s, a physician called Paul of Aegina travelled to Alexandria from Constantinople, probably to collect knowledge to put into the medical encyclopaedia he was writing. This became one of the most accurate and widely diffused medical compendiums, used and cited by scholars across Europe in the succeeding centuries.

It is clear, therefore, that this kind of knowledge was circulating, albeit on a small scale, around parts of the Mediterranean. In order to discover how it reached Salerno, we need to follow the trail that starts in the far south of Italy, in Squillace, at the monastery of Vivarium, where we left Cassiodorus gathering up the remnants of the classical education system. The manuscripts he managed to save and add to his library were, with a few exceptions, the only classical texts on secular subjects available in Western Europe until the eleventh century. Cassiodorus made intellectual study and scribal work an integral part of his monks' daily regime, putting the scrip-

---

* These texts were among those discovered by the Arabs and taken to Baghdad in the ninth century.

torium and book production right at the heart of monastic life. He was an influential player in bringing classical ideals and methods of education into the religious setting, which in turn made the Church the only real option for anyone with intellectual interests and ambitions. The result of this was that "Honour and glory were no longer found in objective, scientific comprehension of natural phenomena, but rather in furthering the aims of the universal Church."[1] Many aspects of ancient scientific and philosophical enquiry were neglected or even destroyed because they did not accord with Christian doctrine. Between 500 and 1100, medicine was the only "profane" knowledge to be continuously studied. The reason for this is perhaps obvious: its urgent, practical utility in the fight against human fragility. And, until Constantine's return to Salerno, the study of medicine in Western Europe was almost entirely defined and facilitated by Cassiodorus.

It was St. Benedict who had first instituted the care of the sick as a fundamental tenet of monastic life. Cassiodorus continued and extended this by including medical texts in his manifesto on education, the *Institutiones divinarum et saecularium litterarum*. This influential book, which was widely in use throughout Europe for centuries, is an encyclopaedic study guide, essentially a list of the texts he expected his monks to become familiar with and discussions of their use. The first part, "Divine," focused on religious texts, but had a section on medicine at the end. The second part of the work, "Secular," examined what would subsequently become the trivium and quadrivium, including maths and astronomy. The medical works he recommended included Dioscorides' *De materia medica* and Latin translations, produced in his own scriptorium, of various works by Galen and Hippocrates. We know that these books were on the shelves at Vivarium in the sixth century, because Cassiodorus explicitly says, "I have left you these books, stored away in the recesses of our library."[2] Cassiodorus thus helped to bring the study and practice of medicine into the realm of the Church, adding to existing traditions of faith healing and visiting shrines. Indeed, it was pilgrims travelling to shrines who would establish another innovation in early medieval medical practice: the foundation of hostels for the care of the sick, travellers and other people in need. These mainly provided food and shelter, but gradually began to offer more specialized medical care as well, fulfilling the ideals of Christian charity. Eventually, and especially in the wake of the Crusades in the eleventh and twelfth centuries, the hostels became hospitals.

In the century after it was written, Cassiodorus' *Institutiones* travelled throughout the Mediterranean world and all over Western Europe, and was a major source for early medieval encyclopaedists. As a catalogue for monastic libraries across the continent, it was one of the most important books in the transmission of knowledge from the ancient world to the medieval one. Moreover, copies of the texts it listed, including works of medicine found in the Vivarium library, were sent to other monasteries, near and far. We cannot trace the precise movements of these manuscripts, but we do know that the medical texts circulating in Southern Italy in the centuries that followed were often the same as those at Vivarium, and some of them certainly ended up in the library at Montecassino. From here, it was a relatively short journey south to Salerno.

St. Benedict's great foundation had had a difficult time since its beginnings in 529. Its strategic position on top of the southernmost point of a spur of the Appenines, towering 500 metres over the Via Latina—the main route from Naples to Rome—made it an obvious target for any armies passing through. The monks fled to Rome when the Lombards sacked the monastery in 581, and did not return until 718. The monastery and church were rebuilt, only to be burned to the ground when the Arabs attacked in 884. This time, the monks headed for nearby Capua, settling there for half a century. When they finally went back to Montecassino, in 949, they found the area "neglected and desolate," but, believing in the sanctity of the location, chosen by St. Benedict, if not by God himself, they persevered in restoring the buildings and encouraging people to come back and live in the surrounding Terra Sancti Benedicti. They even managed to preserve some of the monastery's collections through this traumatic period, an extremely impressive feat, given the violence, destruction and years of exile the monks endured. Two of these, MS Montecassino 69 and MS Montecassino 97, early medical compilations created by the Abbot Bertharius (c.856–883) and based on Cassiodorus' recommendations, have remained in the library to this day.

The medical texts that survived were part of the old Alexandrian curriculum, established in late antiquity, which, by 550, had spread to Italy, Greece and the Byzantine Empire. This consisted of Aristotle's first four books on logic, followed by four Hippocratic treatises. The final section used the "Sixteen Books" of Galen, grouped into seven levels, each focusing on a different aspect of medical practice—for example, anatomy, disease, diagnosis and so forth. Students would attend lectures (often later written down as commentaries) on each of the texts in turn as they progressed through the curriculum. The commentaries on the first four Galenic texts—*De sectis*, *Ars medica*, *De pulsibus* and *Ad glauconem*—were known collectively as *Ad introducendos*, "To beginners." These commentaries, which explained and elucidated the basic principles of Galenic medicine, endured for centuries. They were studied and copied by a handful of people in a couple of places—just enough to ensure their survival through late antiquity and the early Middle Ages. One such man, Agnellus, lived in Ravenna in around 650. He was described as both a "municipal doctor" and a "medical professor,"[3] and he gave lectures on the first four Galenic texts and their commentaries. One of his students, a young man called Simplicius, wrote them down "*ex voce Agnello*," and the lectures survive in a ninth-century manuscript discovered in

Milan, while copies have also been found in Carolingian libraries in Northern Europe.[4]

While many cities in Italy were in decline in the seventh century, Ravenna was having its moment in the sun. Trade was booming in this thriving city, and its 50,000 souls worshipped in splendid new churches decorated with dazzling mosaics. There was considerable scholarly activity, too. Indeed, Ravenna has been described as "the epicentre of the most advanced medical learning in late antiquity."[5] While this statement is true, it gives a misleadingly optimistic impression of the state of that learning. By this point, the medical corpus was a shadow, both in quality and quantity, of the enormous wealth of theory and debate that had prevailed in the ancient world. The dizzying cacophony of ideas and achievements characterizing Hellenistic medical enquiry had been edited down into a rigid, truncated form, ruthlessly shoehorned into the Christian framework of belief. Galen had mutated into Galenism, vanquishing every alternative school of medicine and prioritizing the interpretation of Galenic tracts above all else. This emphasis on theory had the regrettable effect of obscuring the most innovative and useful aspects of the great doctor's achievements—his methods—the observation, empiricism and practical investigation (like dissection, for example) which had taken him so far down the road of scientific progress. The result was that intellectual debate, detailed research and any kind of progress in medicine in the West would remain at a standstill for several centuries. Agnellus of Ravenna and his colleagues made an important contribution to the preservation of medical knowledge, but their achievement should be seen in context.

Ravenna may have flourished in the early Middle Ages, but it was an exception. The general story was one of turbulence and instability. Like much of Southern Italy, the city of Salerno had been a pawn in the wars between the Ostrogoths and the Byzantines in the fifth and sixth centuries, and, when the Lombards arrived in the seventh century, the city was beset by plague and famine. In 774, Charlemagne invaded, deposing the Lombard king. From this moment on, Salerno's fortunes began to improve. As the second city of the Duchy of Benevento, it was fortified in the late eighth century and became the capital of the western half of the Duchy. The city walls were expanded to encompass a new palace and a large residential area with houses, fields and orchards, while a defensive castle behind the city, on Monte Stella, provided a vital lookout post and refuge in times of attack. Landlords encouraged their tenants to grow long-term crops, like vines, hazel and chestnut trees— providing wine and valuable hardwood for trade, bark for tanning and chestnut flour for making into bread or polenta. Culturally and linguistically, the area around Salerno continued to have links with the Greek-speaking Byzantine Empire, not only through trade, but also the Orthodox monasteries whose libraries were home to books

18. A nineteenth-century view of Salerno.

sent from Constantinople, and the possible source of some of the medical texts that found their way to the medical school. In addition to this, Arabs living in Sicily and Southern Italy shared medical ideas and methods with their Christian neighbours.

As a result, a tradition of medical scholarship developed in tandem with the general wealth and success of a city that, by 900, was the most important and prosperous in the whole of Southern Italy. Close ties with neighbouring Amalfi, the first great European maritime power since antiquity, had brought riches to this area of the Italian coast. Amalfitan merchants travelled the length and breadth of the Mediterranean, buying and selling goods; their privileged links with Byzantium and Arab states, notably Egypt, made them major importers of luxury goods from the East, and suppliers of wood, linen and agricultural produce to North Africa. The port of Amalfi itself was tiny and separated from the rest of the country by steep, densely forested mountains, so only the most precious goods, such as exotic textiles, jewels and gold, were imported directly. According to the writer and traveller Benjamin of Tudela, who visited the area in the twelfth century, the Amalfitans were "merchants engaged in trade, who do not sow or reap, because they dwell upon high hills and lofty crags, but buy everything for money."[6] Like their Venetian cousins to the north, the inhospitality of their homeland forced the Amalfitans to be imaginative, to look outwards towards the sea to make their fortunes.* As a result, these enterprising people established colonies in every major port from Rome to Constantinople, via Sicily, North Africa and the Middle East, including a monastery on Mount Athos and settlements inside the Holy City of Jerusalem—decades before the first crusaders arrived in 1095. Salerno's close proximity

---

* The Venetians and the Amalfitans had a lot in common, not least their extensive trading privileges in the Byzantine Empire that played a vital role in creating their mercantile wealth and power.

to Amalfi, broad natural harbour and extensive overland links with the rest of Italy ensured that it was the most important centre for Amalfitan trade; both cities thrived on the profits. As al-Idrisi, the twelfth-century Muslim scholar, noted, Salerno was "a remarkable town, with well-stocked markets and all sorts of goods, in particular wheat and other cereals."[7]

Of all the types of goods being traded and transported, spices were some of the most lucrative and, in relation to this story, the most important. A huge array of roots, berries and plant extracts was imported from distant lands: pepper, ginger, cloves, saffron and cardamom. These spices were, of course, used in cooking, but also in the preparation of medicinal remedies. At this point, Salernitan medicine was primarily grounded in the practical treatment of patients, rather than theoretical expertise. Pharmacology was a significant area of this, especially as practitioners were able to create and study drugs using both local plants and exotic ones supplied by Amalfitan merchants, increasing their repertoire of remedies. They used recipes from Dioscorides' *De materia medica*, copies of which had been circulating in Benevento in various forms since antiquity, supplemented by other local recipes that were added over the years. These collections of remedies were used by all kinds of medical practitioners, and certainly by doctors in Salerno when treating the many patients who arrived there in search of therapy and help.

The medical "school" in Salerno seems to have been well established by the tenth century—its international reputation was such that the Bishop of Verdun travelled there for treatment in the 980s; he must have had enormous faith in the cures on offer to risk the long and hazardous journey from the far north of France. Perhaps he had heard about the hot springs or the benefits of sea bathing in the Gulf of Salerno, both of which were part of the therapeutic regime on offer. There seem to have been several hospitals in the city, many of them infirmaries attached to churches, and, at first,

many doctors were also clerics. As medicine became more profes-
sionalized and academic in the late eleventh and early twelfth centuries, its practitioners were increasingly likely to be laymen—being a
doctor was no longer solely a subsidiary career for churchmen. And,
by the early eleventh century, scholars in Salerno had produced two
major medical texts. The first, by Gariopontus, is the medical compilation *Passionarius Galeni*, which describes diseases and their treatments, starting with the head and moving down the body to the
feet. The second, by Petronellus, adopts the same structure. Both
basically reproduce information from the portions of Galenic and
Hippocratic works and the *De materia medica* that had survived in
the encyclopaedias of late antiquity, but the content is organized in a
more practical form, showing that doctors in Salerno drew on existing theories in the practical treatment of patients.[8] At this point, the
strength of Salernitan medicine lay in its emphasis on the practical—
cures, therapies and diet—but the development of a canon of medical theory was just around the corner.

This canon came to Salerno thanks to Constantine the African.
It will come as no surprise that the facts about his life are few, and
flimsy at that. There are several versions of his biography, which
cheerfully contradict each other on several points.* One of the more
imaginative is by the notoriously unreliable Montecassinan historian
Peter the Deacon, but it does include a list of twenty of Constantine's translations. The most dependable is by Matthaeus F., a mid-
twelfth-century Salernitan doctor, which is appended in the gloss of
Constantine's translation of a book on diet. Out of this confusion,
the following can be ascertained: he was born in what we now call

---

* Other versions of his life are more fantastical, claiming he travelled as far as
India in search of knowledge, and fled Tunisia to escape being murdered by jeal-
ous colleagues.

Tunisia, possibly in or near Kairouan, and came first to Salerno as a merchant, a man who spent his time navigating the sea lanes of the southern Mediterranean, skimming the rocky coast to Cairo. We can imagine him standing on the deck of a wooden ship, eyes narrowed against the glare of the North African sun, scanning the horizon for signs of danger—rocks, reefs, storms or pirates. Like his fellow traders, Constantine also travelled to Sicily—a hazardous journey that forced ships to leave the safety of coastal waters, with their reassuring landmarks and regular ports, and plunge straight out into the open sea, heading east for over 300 kilometres. Today, it is a journey of about ten hours by ferry, but, in the eleventh century, it would have taken about three days, depending on the wind, the weather and the skills of the captain. Constantine would have stopped in Sicily on his way to Salerno. We can imagine the sailors on deck looking out for the first sight of land—the islands of Favignana and Marettimo, or the salt flats on the coast, near Marsala, glittering in the distance. Having completed his business in Palermo, the boat would have set off across the Tyrrhenian Sea, first following the curve of the Sicilian coast, then that of Southern Italy, on their starboard side. On the last leg of the journey, disaster struck. Constantine's ship had just sailed through the Gulf of Policastro and was tacking along the coast when a storm blew up. The ship pitched and rolled, waves crashing over the decks, and, in the tumult, some of the manuscripts were damaged, affecting the quality of the translations Constantine went on to make.

According to the most credible of his biographers, Constantine had spent three years in North Africa gathering these books, before returning to Italy. Together, they represented the full range of medical studies available in this part of the Islamic world, directly descended from the Alexandrian tradition, which had been transmitted to Muslim cities along the North African coast, just as it had to

Italy and Constantinople. Constantine might well have found some of them in the Great Mosque of Kairouan, a thriving intellectual centre where scholars gathered to study and debate under the elegant horseshoe-shaped arches, which were supported by hundreds of ancient columns taken from the ruins of nearby Roman and Greek temples. Medicine was a major preoccupation in the city, which was famous for its talented doctors and, as a result, a good place to find the most up-to-date medical texts. Constantine returned with the best he could find: copies of Hunayn ibn Ishaq's *Isagoge*, al-Majusi's *al-Kamil*, books on urine, fevers and diet by the Jewish scholar Isaac Judaeus (d. 979), the *al-Hawi* of al-Razi, a medical guide for travellers by the Kairouanese doctor al-Jazzar (895–979), and a treatise on sexual intercourse, called *De coitu*. He would translate many of these into Latin himself. He also brought a book on melancholy, which he dolefully noted was "very prevalent in these regions."[9] These texts became the basis of the medical curriculum throughout Europe. They would remain influential for centuries, with printed editions produced in Lyon in 1515, and Basel in 1536. Constantine did not leave us any idea as to why he did this extraordinary thing, what impelled him to devote his life to bringing medical knowledge to a continent he hardly knew; we can only guess at his motives.

After Constantine settled in Salerno, he got to know the city's archbishop, Alfano, who shared his passion for medicine. Like many of the scholars we have met on this journey, Alfano was an extraordinary character—a gifted scholar, with eclectic intellectual interests that encompassed classical literature, architecture, theology and science. His family was wealthy and influential, ensuring he benefited from the best education on offer. Alfano was an accomplished poet, he oversaw the building of a new cathedral, and he was also a talented physician and a leading proponent of the medical school. Having perfected his Greek while visiting Constantinople as a young man, he produced a translation of *On the Nature of Man*, a text he probably picked up on his travels in the East, which included a pilgrimage to Jerusalem. This was a wide-ranging philosophical work, written in the fourth century by Nemesius, Bishop of Emesa, who had been especially influenced by the writings of Galen, Plato and Aristotle. In producing his translation, Alfano began the process of creating a new technical Latin vocabulary with which to express the complex scientific and philosophical ideas in Nemesius' treatise. At around the same time, Constantine was translating the *Isagoge*, and the two men may well have collaborated; they would certainly have had much to talk about. Alfano, who had the wealth and power of both his family and the Church at his fingertips, supported Constantine financially, paying for his translation of al-Majusi's huge medical compendium, the *Pantegni*. In turn, concerned about his friend's health problems, Constantine produced a compilation of advice on stomach complaints. Together, they set about revolutionizing the study of medicine in Salerno and beyond, developing new Latin terms to bring the wealth of Graeco-Arabic knowledge to Western Europe.

Alfano had many influential friends, including Desiderius, Abbot of Montecassino. The pair had met when Desiderius visited Salerno in the 1050s for medical treatment; they bonded over shared intellectual interests and became close. Desiderius persuaded Alfano to return to Montecassino with him, to study there for a while. In all likelihood, it was Alfano who now suggested to Constantine that he should move there, too. He couldn't have chosen a better time; Montecassino was basking in a golden age, the most influential, prestigious religious foundation in all Europe. The chance to work in the busy scriptorium, home to the distinctive Cassinese script, surrounded by other scholars and with all the practical paraphernalia, like parchment, ink and pens, freely available, to say nothing of an army of scribes to assist him, must have been irresistible. Most important of all, he would be able to study the library's medical manuscripts and use them as context when preparing his own translations for Latin readers. It is also possible that Constantine had religious reasons for wanting to become part of a monastic community. We do not know whether he converted from Islam at some point during his time in Italy, or whether he was, in fact, born a Christian—there were several Christian communities in North Africa at that time.

The journey from Salerno to Montecassino would have taken several days, and Constantine must have set off up the old Roman Via Popilia, which skirted Naples on its way to Capua, and, from there, taken the Via Latina northwards. He would have seen the monastery from miles away, on its rocky hilltop perch overlooking the farms and fortified villages of the Terra Sancti Benedicti in the Liri Valley below. After a gruelling climb up the hill, he would have entered the gates of the enormous monastery complex and passed the site of the new basilica church that was under construction, being decorated with elaborate mosaics, textiles and jewels by Byzantine craftsmen. The massive bronze doors, specially cast in Constantinople with silver-inlaid panels, may well have already been installed.

At some point after his arrival, Constantine was taken to meet the author of all this splendour, Desiderius. He presented the Abbot with letters of recommendation from Alfano, and a copy of his recent translation of the *Isagoge*. While at Montecassino, Constantine completed his biggest project, the *Pantegni*, and dedicated it to Desiderius. As the first comprehensive medical text in Latin, it was hugely important, but also controversial. Although much of Constantine's text is based on the *Kitab Kamil—The Complete Book of the Medical Art*, by Ali ibn al-Abbas al-Majusi (d. *c.*982), it is far from a faithful translation; it is truncated in some places and expanded with alternative sources in others. Constantine makes no mention of al-Majusi, or indeed the authors of the other sources he included, seemingly presenting it as an original work.* He also edited out all al-Majusi's references to earlier Arabic scholars, instead prefacing the translation with a list of the sixteen books in the Alexandrian curriculum—effectively erasing the Arab contribution and emphasizing the importance of Galen. Understandably, this has led to accusations of plagiarism by modern historians, but it looks as if Constantine was deliberately attempting to conceal the Arabic origins of the text in order to maximize its potential in Europe, rather than trying to claim ownership himself. Recent political events, especially in Southern Italy where "Saracen" raids had caused considerable death and destruction, meant that the general attitude towards Muslims was not especially receptive. On the other hand, the medical knowledge in Italy at the time came from the Hellenistic tradition, so Constantine was perhaps also ensuring that his work was in accord with prevailing ideas. Strangely, however, he generally chose to translate Arabic versions of ancient texts, rather than the Greek originals, so he must have thought they were superior, even though he then obscured their authors and emphasized their Greekness.† This bewildering tangle of cultural priorities shows how complex the relationship between Europe and Islam was at the time.

Constantine names only two of the writers whose work he translated: the Jewish doctor Isaac Judaeus, who in turn took much of his information from Galen, and Hunayn ibn Ishaq, whose name he Latinized to Iohannitius. He also Hellenized many of the titles of

---

* "Haly Abbas," as al-Majusi became known in Western Europe, was a mysterious but brilliant figure from Persia, one of the three greatest (along with al-Razi and Avicenna) physicians of the Eastern Islamic Empire and, judging by his name, from a Zoroastrian family. The *Kitab Kamil* had been extensively copied and translated in Arabic, Hebrew and Urdu by the time Constantine produced his Latin version. It was the most important medical text in the Arab world until Avicenna's *Canon* appeared.

† For example, in the *Isagoge*, he translated Hunayn's version of Galen's *Art of Medicine* (*Tegni*), as opposed to the original.

the works he translated and adapted their contents for a Latin audience. This is especially true of the *Pantegni* (Greek for "all arts"), which is based on the fundamental structure of al-Majusi's *Kitab Kamil*, but misses out huge sections of the original in favour of material from other treatises and includes a literary discussion of medicine, making it a related but quite different work. This was partly necessity—the manuscript was damaged during the storm on the journey from Africa to Italy, so was incomplete and missing the last few sections—but also because Constantine was engaged in creating a practical curriculum to educate young doctors, rather than producing faithful renditions of the source texts. Incidentally, the *Pantegni* is also full of mistakes and confused meanings, but was widely used as a manual on the fundamentals of medicine nonetheless. Incredibly, a manuscript copy of the *Pantegni*, produced in the Montecassino scriptorium and supervised by Constantine himself in the late eleventh century, has survived and is now in the Koninklijke Bibliotheek in The Hague. The *Pantegni* was read and quoted by many European scholars in later centuries, especially in works of natural philosophy. Daniel of Morley got hold of a copy, possibly while he was in Paris, and, as we will see, Adelard of Bath also quoted it extensively. In addition to being dispersed across Europe, the *Pantegni* became a standard text on the Salernitan curriculum and had huge influence, especially on the study of anatomy—Constantine included two books on the subject that were not part of al-Majusi's original work, but derived from Galenic texts.

These were the first classical texts on anatomy to be available in Europe. Those on offer to students in Salerno were extremely limited—the Church frowned upon the study of anatomy, and this aspect of medicine had effectively been edited out of the curriculum. However, this was beginning to change and, by the time Constantine's translations appeared, masters were demonstrating anatomy to their students by dissecting pigs, something that would soon become an annual event.* A physician called Master Copho wrote the earliest record of such a dissection, and his *De modo medendi* was the first—and the most primitive—of a series of four texts on the subject, which together became known as the *Anatomia porci*. The second was much more detailed and drew heavily on Galenic anatomy as translated in Constantine's *Pantegni*. Several versions of these manuscripts survive and they were still used by students as an introduction to anatomy, even after Vesalius' work on human cadavers transformed the subject in the sixteenth century. These anatomical demonstrations show that Salernitan medics were looking at nature for themselves, rather than blindly relying on the limited information that had come down to them from Byzantium and Alexandria. They were returning to ancient, specifically Galenic, methods of study. The *Anatomia porci* are essentially written records of the lectures given by Salernitan masters as they cut up pig carcasses in front of a classroom full of students. By the mid twelfth century, then, it is clear that doctors studied anatomy as part of their training. But it is striking that, in the long history of anatomy, there was no formal human dissection between 150 BC and AD 1315, when the first cadaver was publicly dissected at the University of Bologna. In Constantine's time, however, autopsies to discover the cause of death had become almost commonplace, at least in Italy. And, in 1231, the Holy Roman Emperor, Frederick II, decreed that human corpses should be dissected publicly at least once every five years. Whether or not this actually happened is another matter. Interestingly, society's natural aversion to such a practice was moderated by the need to prepare the corpses of fallen crusaders, so that their hearts could be returned home for burial. Using animals like pigs and apes could only take the clinician so far, as Galen and the Salernitan anatomists showed. The only way to discover the secrets of the human body and the dis-

---

* Galen also used pigs, which are anatomically similar to humans, for dissection.

eases that consume it is to open it up and look inside, but this would not become the norm until the Renaissance.

While the *Pantegni* would have lasting significance, and was soon in circulation far beyond Salerno, Constantine's translation of the *Isagoge* would be even more influential. Especially when, in the thirteenth century, it was combined with new translations of Hippocrates' *Aphorisms* and *Prognostics*, with their Galenic commentaries and two Byzantine treatises, one on urine and one on the pulse.* This collection of texts, known as the *Articella*, formed the medical curriculum of Western Europe for the next 500 years. It built on existing medical tradition, adding detailed information and shaping the whole into a properly structured discipline. The *Isagoge* was the cornerstone of the *Articella*, the first Arabic medical text translated into Latin. Translated by Constantine while he was living in Salerno, it was a version of Galen's *Art of Medicine*, produced by Hunayn ibn Ishaq and titled the *Masa'il fi-Tibb*—a direct line of transmission, but, like the *Pantegni*, a loose anthology, rather than a comprehensive translation. Again, Constantine was ruthless in his treatment of the Arabic source. He removed the dialogue—Hunayn's text was based on questions and answers—replacing it with a catalogue structure more suitable for lecturing. He also mistranslated and edited some sections so drastically that they no longer made any sense. This might have mattered to a culture with a sophisticated tradition of medical study, but it certainly didn't hinder the *Isagoge*'s progress in Europe.[10] Inaccurate information was better than no information at all, and copies moved fast from city to city; scholars in Chartres were already writing commentaries on it by 1150, and, by 1270, it had been adopted as the basis of the medical schools at the universities of Paris and Naples, among others.

* In ancient medicine, doctors used patients' urine and pulse to diagnose complaints and assess where the humoral imbalance lay.

While Constantine was busy with his ambitious translation project, Alfano was dealing with seismic changes in Salerno's political climate—changes that would have profound consequences for the transmission of knowledge to the rest of Western Europe. It is not clear when Constantine left the city for the peace of Montecassino, but it might well have been as a result of a new force making itself felt—the Normans, who conquered Salerno in 1077, after a brutal siege. It was a busy time for these ambitious, energetic descendants of the Viking raiders who had settled in northern France a couple of centuries earlier. In the 1060s, when William the Conqueror began to look hungrily northwards, over the Channel, towards Britain, many of his fellow Normans had already gone south to fight as mercenaries for the Lombards, who were trying to win independence from the Byzantines in Southern Italy. The Norman knights were skilled fighters and proved very useful to their employers, helping them repel the Byzantines and win the campaign. The Lombards rewarded them with lands, and they began making themselves at home in Benevento and Calabria—settling, intermarrying, inserting themselves into local society and forming their own power bases. And, in a situation that has played out many times in history, the saviours became the aggressors, and eventually the conquerors. It wasn't long before Norman knights were charged with protecting the Pope himself, who, in 1059, rewarded their leader, Robert Guiscard, with the duchies of Apulia, Calabria and Sicily. This grant transformed the Normans' influence; from then on, it was abundantly clear that they were not only here to stay, but here to rule. Guiscard, the sixth of twelve sons fathered by Tancred de Hauteville, a minor Norman nobleman, was nicknamed "the cunning" or "the fox."*

---

* Tancred was married twice; his first wife gave him five sons and one daughter, his second wife seven sons and at least one daughter. Robert was the eldest of the second family. Most of the brothers lived in Southern Italy, where they fought one another relentlessly for power and land.

The Byzantine historian (and princess) Anna Komnene described him thus:

> an overbearing character and a thoroughly villainous mind; he
> was a brave fighter, very cunning in his assaults on the wealth and
> power of great men; in achieving his aims absolutely inexorable,
> diverting criticism by incontrovertible argument. He was a man
> of immense stature, surpassing even the most powerful of men; he
> had a ruddy complexion, fair hair, broad shoulders, eyes that all
> but shot out sparks of fire . . . Robert's bellow, so they say, put tens
> of thousands to flight.[11]

In the light of this, it's no surprise that Guiscard threw himself into the task of subduing and uniting Southern Italy, in addition to invading Sicily, along with his brother, Roger de Hauteville. Having seized Salerno, the last remaining independent city on the Southern Italian mainland, Robert made it his capital and began building a new cathedral, with the help of Alfano, who had supported his invasion.

The Normans would dominate Southern Italy for the next hundred years, dramatically shifting the balance of power in Europe, and creating strong links between the south of the continent and their northern homelands, with which they retained close contact. Guiscard and his brother Roger (now ruler of Sicily, but still nominally subordinate to Robert) appointed fellow Normans from England and France to positions in the Italian and Sicilian churches, while southern clerics went to study in monasteries like Bec, in Normandy. This was mirrored in the royal courts, with John of Salisbury travelling to Southern Italy more than once to study Greek and philosophy, and other scholars from England going to study in Palermo. And, in Salerno, there are several English names in manuscript lists of doctors who studied there.[12] Naturally, trade links had also opened up; there are records of at least one English merchant in Salerno, and a merchant from Brindisi visited the Canterbury shrine of St. Thomas Becket, who had recently been added to the Southern Italian calendar of saints. These connections ensured the flow of books, too. Constantine's translations were soon bound together in compilations for easy use, travelling north up the Italian peninsula, over the Alps, through the wild forests of France and even across the choppy grey sea to England. Scribes, hunched over desks in the flickering yellow of tallow candle flames, scratched out copies that were placed in library cupboards and used as textbooks for teaching medicine at universities.

A couple of decades after Guiscard's triumphal entry into Salerno, the Norman position was enhanced as thousands of men from Northern Europe set off to the Holy Land on the First Crusade, with Sicily and Southern Italy natural stopping-off points on the journey east. There had been no travel on this scale since the days of the Pax Romana. The Crusade was part of the same tectonic shift that had seen Christian power move southwards into traditionally Muslim areas in Spain, a shift that had a profound—and irreversible—effect on culture, politics and society. The European world was opening up, and flexing its muscles. The Crusade also changed Salerno. Hundreds of wounded men returning from the Holy Land were treated there on their long journeys home, enhancing the city's reputation as a centre of medical excellence.

Constantine died sometime in the last years of the eleventh century, at Montecassino. After his death, his pupils at the monastery, John Afflacius and Azo, continued his work in the scriptorium, helped to promote, in the wider world, both his translations and those they produced themselves, and wrote their own medical treatises. Afflacius finished translating the *Pantegni* and sent copies of it, the *Isagoge* and other translations to Salerno, where, early in the twelfth century, scholars began producing commentaries and

19. Robert II of Normandy being treated in Salerno for injuries he sustained fighting in the First Crusade.

teaching guides on these texts. The *Articella* was developing into an organized set of textbooks designed for teaching; as time went on, it was added to and transmitted to other centres of learning. In 1161, for example, the Bishop of Hildesheim had twenty-six medical text books in his library; the majority of them were Constantine's translations. At the same time, interest in the work of Aristotle was rekindled, largely due to the work of one Urso, a scholar from Salerno. Aristotle's insistence on the proper observation of the natural world, on experimentation and critical thinking would have a profound influence on almost all aspects of intellectual life, and the teaching of medicine was no exception. As universities sprang up throughout Europe, the study of medicine was gradually elevated to a theoretical, academic subject, part of the liberal arts curriculum, integrated with Aristotelian natural philosophy and based on sound textual foundations.

In Salerno, and elsewhere, the masters of medicine, following Galen, began to focus on the causes of disease and used their discoveries to decide on the type of therapy they should prescribe. It was an entirely new way of looking at illness. Previously, disease had been attributed to the wrath of God, or to being possessed by evil spirits—ideas that naturally discouraged rational observation and investigation. Medicine was no longer just *medicina*—a mechanical art, limited to practice—now, it could take its place alongside the other natural sciences. As medical literature grew to include analytical textbooks—for teaching and study, rather than the simple reference manuals of the previous period—it became a systematized, uniform discipline, based on a universally accepted set of authorities and ideas—ideas that would change as new texts were translated and new discoveries were made. As time went on, some of Constantine's other translations began to be included in the *Articella*, and medical study grew and flourished. This in turn affected the status of medical practitioners, who were increasingly differentiated between the learned and the unlearned. A formal system of qualification developed, with students having to follow the course laid down in the *Articella*. Studying medicine took several years, and the number of students who completed the course was relatively small. By the mid thirteenth century, students had to study logic for three years before they were allowed to begin the five-year medical curriculum. And, after that, they had to complete a year of practical study with a qualified physician. Only then could they take the examinations and—if they passed—receive their formal licence to practise. Doctors did become slightly more numerous, but only a tiny, urban elite had access to their services; the majority of people continued to rely on herbal remedies and advice from local, untutored practitioners.

Collections of these remedies, originally based on Dioscorides' *De materia medica*, had been in circulation for centuries, added to and edited to suit the needs of whoever was using them. In twelfth-century Salerno, two main versions, usually found together in the same manuscript, were in use: the *Circa instans*, which was in the *Materia medica* tradition and focused on "simples," or remedies, made from just one substance, and the *Antidotarium Nicolai*, which was an amorphous collection of recipes for compound medicines

20. A page from a manuscript copy of the popular *Circa instans* with a drawing of a turpentine plant.

with several ingredients. The *Circa instans* describes each plant, mineral, root or fungus in detail, listing its properties according to Galen's system of degrees and elements. For example, cardamom "is hot and dry in the second degree," which meant that it was good for treating people who suffered from the opposite (cold and wet). This system enabled physicians to prescribe the correct amounts of the substance to rebalance the patient's four humours. A Salernitan scholar called Mattheus Platearius wrote the most successful version of the *Circa instans*, which was translated into all major European languages. In some places, apothecaries were required by law to have a copy of it in their shops, so it is no wonder that so many manuscripts have survived.

The *Antidotarium* lists detailed recipes for a huge variety of medicines, half of which come directly from the *Pantegni*. Here is the recipe for a "soporific sponge," used as an anaesthetic and a sleeping draught:

> Take one ounce of opium from Thebes, then one ounce each of
> juice of jusquiam [henbane, hyosciamus], of unripe mulberry,
> of blackberry, of seeds of lettuce, of hemlock, of poppy, of
> mandragora, of arboreal ivy. Put all these in a vessel, together
> with a sponge just taken from the sea so that it has never been in
> contact with fresh water. Expose [the vessel] to the sun during the
> dog days until everything is consumed. When you want to use
> it, moisten the sponge lightly with hot water and apply it to the
> patient's nostrils, who will quickly fall asleep.[13]

This exotic cocktail could have had all kinds of side effects, but sleep was unlikely to be one of them. It is, however, reminiscent of al-Zahrawi's rather more successful recipe for anaesthesia: a sponge soaked in cannabis and opium. Of all the remedies in the *Antidotarium*, the one that makes the grandest claims is "The Great Theriac of Galen," a magical, cure-all potion, supposedly created by the man himself and apparently effective against a dizzying range of conditions, including apoplexy, epilepsy, migraine, stomach ache, dropsy, asthma, colic, leprosy, smallpox, chills and all poisons, snake and reptile bites. The list of ingredients is so long and so complicated that it is hard to believe anyone ever managed to successfully produce it, but, if they did and if the "Great Theriac" lived up to its reputation, they could have single-handedly cured the entire local population.[14]

These medicines demanded a huge variety of plants, and specialist physic gardens had long been a feature of monastic establishments, cultivated alongside the kitchen gardens that produced food for the monks. Local, secular healers must have also had their own herb gardens, and twelfth-century Salerno was full of small plots and orchards for growing fruit, vegetables and herbs. Houses were relatively small, wooden structures that could easily be taken down and moved elsewhere, and most people grew their own food and kept pigs, chickens and geese, if they could afford them. Apothecaries and pharmacists would also have had their own specialist herb gardens to supply the plants to make their remedies, supplemented by exotic ingredients purchased from the city's merchants.

There had always been plenty of wise women dispensing advice and remedies to their local community, but, in Southern Italy, there were also—remarkably—learned female doctors, trained and educated in Salerno and Naples. Tragically, this enlightened aspect of Salernitan medicine did not catch on elsewhere, and, with a very few exceptions, women had to wait until the twentieth century before being able to study and practise medicine in significant numbers. These medieval women were especially skilled in gynaecology, obstetrics and female health; their combined knowledge was expounded in the twelfth century in a series of three texts known as the *Trotula*. The origins of the *Trotula* are unclear, but a woman from Salerno called Trota or Trocta may well have been involved with their creation and thus given them her name. The texts cover a range of subjects—pregnancy, childbirth, even cosmetics—and, drawing on Arabic sources translated by Constantine, they integrated gynaecology into the Galenic framework of humours, and thus into the emerging academic curriculum. Drawn from newly translated texts, but grounded in years of practical knowledge of midwifery and women's health, the *Trotula* was used by doctors all over Europe. Most of them were, of course, men, who were probably relieved to have some proper insight into the mysterious workings of the female body.

Despite the visionary admittance of women to its medical school, by the early thirteenth century Salerno had lost its dominant position in European medicine. Cities like Bologna, Montpellier and Padua rose to take its place, their curriculums based on Constantine's translations and books by Salernitan medical scholars. Scholars still came to study in Salerno, and they took the textbooks of the *Articella* back with them, ensuring that the city remained an important gateway to medicine in medieval Europe. But its medical school declined, overshadowed by that of its neighbour, Naples, with its growing university and promotion to the capital of the Kingdom of Sicily and Southern Italy.

The previous capital of the region had been Palermo, an elegant city on the north-western coast of Sicily. This had been the Norman rulers' major power base throughout the twelfth century, home to their glittering, cosmopolitan court. As we shall see, this court's diplomatic connections with Constantinople opened up new avenues of cultural exchange as envoys brought books back to Sicily, echoing what had happened in Baghdad and Córdoba. But Norman culture was predominantly Latin, so these Greek texts were translated directly, rather than through the medium of Arabic, as they had been in Toledo and Salerno. This new strand of transmission would go on to play a profound role during the Renaissance, when humanist scholars sought out and revered ancient texts in the original Greek, but it had its beginnings in twelfth-century Palermo.

SEVEN

Palermo

The first of these towns is Palermo, a city that is both remarkable for its grandeur and most illustrious for its importance . . .

This town is on the coast; it has the sea to the east, and is surrounded by high and massive mountains . . .

The town is endowed with magnificent buildings, which welcome travellers and flaunt the beauty of their construction, the skill of their design and their marvellous originality. Along the central street one finds fortified palaces, high and noble residences, and many mosques, hostels, bath houses, and the warehouses of great merchants.

The town is traversed on every side by watercourses and springs; fruits grow in abundance; its buildings are so beautiful that it is impossible for the pen to describe them or for the mind to imagine them; everything is a real seduction for the eye.

—Al-Idrisi, *The Book of Roger*

Where are you hurrying off to? To where do you wish to return?

In Sicily you have the Syracusan and Argolian libraries; there is no lack of Latin philosophy.

You have the *Mechanica* of the philosopher Hero [of Alexandria] to hand . . . the *Optica* of Euclid . . . the *Apodictice* of Aristotle on the first principles of knowledge . . . the *Philosophica* of Anaxagoras, of Aristotle, of Themistios, of Plutarch and of other famous philosophers are [also] in your hands . . . you may consult a good work devoted to the study of medicine, and I can also offer you theological, mathematical and meteorological tracts [*theoreumata*].

—Henricus Aristippus,
dedicatory letter to the *Phaedo*, 1160

IN 1160, a young man was studying medicine in Salerno. We do not know his name or where he came from, but we do know that he was particularly interested in astronomy. So interested, in fact, that when he found out a manuscript of Ptolemy's *Almagest* had been taken to Sicily, he deserted his studies and set off to find it. The book had been brought from Constantinople, apparently from the library of Emperor Manuel Comnenus himself, by Henricus Aristippus, a scholar, archdeacon and, most importantly, chief advisor to King William I of Sicily, who had sent him to Constantinople to negotiate a peace treaty. The talks went well and Aristippus seems to have impressed the Byzantines, which gave him the opportunity, like many scholar-diplomats before him, to get hold of some manuscripts while he was in the ancient city. How the young medic in Salerno discovered this piece of news is a mystery, but it does reveal something about the close connections between Sicily and Salerno at this time, and also shows that the fame of Ptolemy's great masterpiece had reached Southern Italy. The parallels with Gerard of Cremona, who, just a few years earlier, had travelled to Spain in search of the very same book, are striking. There is no doubt that European scholars were becoming aware of the wealth of classical and Arabic science, and resolving to get their hands on it. The Salernitan scholar's journey to Sicily was considerably shorter than Gerard's odyssey across the Mediterranean, but it was nonetheless fraught with dangers—"the terrors of Scylla and Charybdis," as he puts it— the legendary winds and whirlpools that made the narrow crossing from Southern Italy to Sicily so difficult. Once safe on Sicilian soil, he headed for Catania, where Aristippus was archdeacon. However, the great man was not, as might perhaps have been expected, at his desk in an elegant palace, nor was he celebrating Mass at the altar of his cathedral. The young scholar was forced to travel onwards, "to navigate the fiery rivers of Etna" and climb to the summit, where he finally found Aristippus, on the crater's edge, studying volcanic activity.[1]

This anonymous scholar is just one of many people who travelled from Salerno to Sicily in the eleventh and twelfth centuries. The two places were closely linked, not least by the fact that they were both ruled by the de Hauteville dynasty. In 1061, having subdued most of Southern Italy, Robert Guiscard set off to conquer Sicily with Roger, one of his many younger brothers, in tow. The de Hautevilles were intimidatingly fecund. Coming from a family of twelve sons and at least two daughters, it is hardly surprising that so many of them left Normandy in search of fortune elsewhere. Even a seriously prosperous noble would struggle to provide for so many heirs, and their father, Tancred, was far from wealthy. As Amatus, one of the great chroniclers of the age, bluntly explained, "Abandoning little in order to acquire much, these people departed, but they did not follow the custom of many who go through the world placing themselves in the service of others; rather like the ancient warriors they desired to have all people under their rule and domination."[2] Several of them came to Southern Italy and made it their own personal playground, playing out their relentless sibling rivalries with absolutely no regard for anyone else. We can only imagine how terrifying it must have been to be caught up in the maelstrom of violence as the de Hauteville brothers trampled across Apulia and Calabria, making and breaking alliances with local rulers, the papacy, the Byzantines and, most of all, each other.

Both Robert and Roger continued the family tradition of fertility. Robert had four sons and seven daughters, while Roger married three times, producing no fewer than seventeen children within wedlock, and probably a few without. De Hauteville daughters were used as pawns in an ambitious dynastic strategy to propel the family ever upwards. One of Roger I's daughters married Conrad, son of the Holy Roman Emperor, Henry IV; another married Coloman, the King of Hungary. Both girls brought their husbands large dowries. When marriage alliances weren't an option, the brothers used a combination of brute force and cunning. Opportunistic, violent and irrepressible like the rest of their clan, they dedicated their considerable energies to conquering Sicily, which was at the time in the hands of several feuding Muslim warlords.

This was not the first time Sicily had been invaded, nor was it to be the last. As the largest and most strategically important island in the Mediterranean, it had been at the top of every empire's invasion list since memory began. When the de Hautevilles arrived, they encountered a population made up of communities of Jews, Greeks,

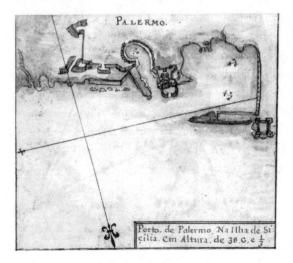

21. A Portuguese map of Palermo harbour.

Muslims and even the odd Latin Christian. The area around Messina, where they first landed, had been predominantly settled by Greeks. Sicily's links with Greece began in 750 BC, when Greek settlers came to the island and began to establish colonies there, assimilating with the indigenous population. They founded important settlements, Greek culture and religion flourished, and the island became part of what was called Magna Graecia. Then, as the Roman Empire began to expand beyond the borders of Italy, Sicily was an obvious target, and, in 242 BC, it became the first Roman province outside the mainland. This was not surprising given the beauty and fertility of the island, where vines and olive trees, introduced by the Greeks, flourished. The Romans put the rich, volcanic Sicilian soil to use by growing wheat in such quantities that Cicero, quoting Cato the Elder, called the island, "the Republic's granary, the wet nurse of the Roman people."[3] Wealthy Romans built opulent villas, where they relaxed and took their leisure, hunting specially imported exotic beasts, drinking the local wine and carousing with beautiful maidens. These festivities are celebrated in fabulous mosaics, some of the best preserved in the world, at the Villa del Casale, at Piazza Armerina, in central Sicily. In the centuries after the collapse of the Roman Empire, Sicily was invaded many times, underlining the island's strategic importance at the centre of the Mediterranean, close to Southern Italy and with easy access to North Africa, Spain and the Middle East. Both the Vandals and the Ostrogoths took control, albeit briefly, before they were defeated by the Byzantines, who reintroduced Greek culture and language. They even moved the capital of the Eastern Roman Empire from Constantinople to Syracuse for a while.

After the Arabs invaded in the ninth century, it took several decades for them to completely conquer Sicily. And, even once they had done so, their characteristically tolerant attitude to other faiths meant that Jewish and Christian Sicilians were allowed to live and worship in peace, so long as they paid the *jizya*, the "non-Muslim" tax. The new rulers brought new crops with them, and sophisticated irrigation systems that extended the growing seasons and transformed Sicilian agriculture. The island exported wheat and rock salt, vital for the preservation of food, to North Africa and beyond. Muslim immigrants (Arabs, Berbers and other tribes) arrived from North Africa and settled happily in the lush landscape, but these communities fought both among themselves and with the Byzantine Greeks, who lived in the north-east. By the early eleventh century, the island had split into a series of warring provinces led by local warlords. It was ripe for invasion.

The Christian reconquest began when the Lombard lords of Southern Italy, keen to win Sicily back from the Muslims, recruited Norman mercenaries to attack them. So, when Robert and Roger landed with their force of a few hundred knights, they were not the first of their kind, nor even the first of their siblings, to do so. In 1138, their older brothers, William "Iron Arm" and Drogo, had made such a nuisance of themselves in Apulia that the Prince of Salerno, in an attempt to get them out of the way, sent them to Sicily to help the Byzantines fight the Arabs. This backfired when, unhappy with their share of booty, the Norman troublemakers returned to the mainland and settled in the Byzantine area around Melfi, where they built a vast stone fortress. Initially, at least, Robert and Roger had been invited to fight in Sicily on behalf of a local leader trying to dominate his rivals, but, by 1091, they had won the island for themselves. By this time, Robert had returned to his dominions on the mainland, where he was now Duke of Apulia, leaving Roger in charge of the campaign to rid the island of Arab rule, which took several decades to complete. Even after Noto, the last Muslim stronghold, fell in 1091, controlling the divergent ethnic cocktail that made up Sicilian society took constant vigilance—and an iron fist. According to the chronicler Hugo Falcandus, Roger de

Hauteville "made efforts to administer justice in its full rigour,"[4] while another writer described him as appearing "so terrible that even the mountains trembled before his countenance."[5] This was necessary, "for there was no other way in which the savagery of a rebellious people could have been suppressed, or the daring of traitors restrained."[6] Roger needed to establish his rule and impose stability. As both foreigners and minor nobles, the de Hautevilles had to rely on brute force, opportunism and wily negotiating to maintain their hold on power. Under their control, Sicily became one of the wealthiest states in Europe.

This success was down not only to the de Hautevilles' strength of arms, but also to their rampant ambition. Roger, now Count of Sicily, but nominally still a vassal of his brother Robert (not a position he relished), brazenly helped himself to every Byzantine or Arabic idea he fancied to form his government and fashion his image. His administration was organized on existing Byzantine traditions,* he adopted many Arabic customs and audaciously ordained that, when he died, he was to be entombed in an ancient Roman sarcophagus with a porphyry marble canopy—an honour hitherto reserved for Byzantine emperors.[7] The message was clear: the de Hautevilles were major players on the world stage, adorned with all the pomp and ceremony that entailed. Under their influence, the royal culture of the Arab and Byzantine worlds spread through Europe. From now on, lavish ostentation would be inextricably linked with majesty and the expression of princely power. Throughout the twelfth century, the dynasty celebrated and thanked God for its success by founding three new cathedrals—in Palermo (where the old basilica had become a mosque under Muslim rule), at Monreale and in Cefalù—while a host of newly built churches and monasteries pro-

---

* These included the way taxation was organized, aspects of the legal system and methods of listing serfs.

claimed the return of Christianity to the island. Dilapidated city walls and defences were restored, security was imposed and communities of merchants began to arrive and settle in their own quarters of Sicily's towns and cities, filling the void left by the Muslim and Jewish merchants who had fled during the chaos of the invasion campaign.[8] The Amalfitans, always quick to capitalize on an opportunity, established their own street in Palermo, lined with shops, and even built their own church, dedicated to Sant'Andrea. Palermo flourished, markets grew, "mansions like lofty castles with towers hidden in the skies" rose up and gardens bloomed around them, nourished by a complex hydraulic system that drew water up from underground springs.[9]

Roger was succeeded by his son, Roger II, who inherited many of the traits that had taken his father and uncles from penury in Normandy to glory on the shores of the Mediterranean. But he had something they lacked: education. Roger I died when his son was still a child, leaving his young wife, Adelaide, as regent. A formidable woman, Adelaide had married Roger when he was sixty and she was only fifteen. She oversaw her son's schooling, and ensured that his thirteen older siblings by her husband's first two wives were firmly excluded from the succession. Roger II spent his early years in Messina, on Sicily's Greek-dominated eastern coast, where he was educated by Christodoulos, a Sicilian of Greek-Byzantine descent, who was also Adelaide's chief advisor. Christodoulos instilled in his young pupil a love of learning and culture that would last for the rest of his life. In around 1111, when he was sixteen years old, the young Duke's court moved to Palermo, a city imbued with Arab influences, which opened his eyes to the rich variety of Sicilian culture.

Roger II continued many of his father's policies. He was tolerant of other faiths and styled himself as the protector of all the peoples he ruled. His main concern was keeping control of his kingdom and ensuring peace and stability wherever possible. It was a constant struggle. The open-minded atmosphere that prevailed at court was not reflected in the fields, villages and small country towns, where people were rarely well integrated. In rural Sicily, Christians and Muslims lived quite separately, in different areas and in different settlements, fostering unease and antagonism, which often spilled over into violence. Raids were common, especially by the recently arrived settlers from the mainland, who were keen to extend their territories at the expense of local Muslim communities. Keeping local warlords in check continued to be a major preoccupation for the Norman administration. It was quite a different story in the sophisticated world of the court, where the brightest minds were welcome, regardless of faith or race, and where merchants from across the known world traded and tricked one another in every language under the sun, and diplomats arrived from afar to advance their countries' interests. And, in the great cities, like Palermo, even though communities tended to group together based on faith, settling in particular quarters, people lived in such close proximity to one another that it encouraged friendly relations and mutual cooperation.

This cultural openness was instituted by Roger himself. No doubt it was, in part, a pragmatic solution to the challenges he faced as the head of a tiny, foreign elite, ruling over a very diverse population, but Roger was also genuinely interested in other cultures and celebrated the traditions of his Muslim-Arab and Greek-Byzantine subjects.* His love of Arabic customs is well attested. He sat in state beneath a bejewelled parasol, a gift from the Fatimid Caliph, with banner carriers and shield bearers in attendance. A splendid vermilion silk mantle survives in a museum in Vienna. Made around 1134 by Muslim artisans in the palace silk workshop, the mantle is adorned with two lions, each attacking a camel, flanking a palm tree in the centre, all richly embroidered in gold thread studded with

---

* At this point, Muslims still made up the majority of the population of Sicily.

garnets, rubies and pearls. An inscription in Arabic, stitched around the hem, explains where and when it was made.[10] Roger's court was famously trilingual; he used Greek, Latin and Arabic epithets, often signing himself "Basileus" instead of "Rex," calling himself "the defender of Christianity," on one hand, and "powerful through the grace of Allah," on the other.[11] The Greek, Latin and Arabic scholars Roger employed wrote official documents in all three languages, the most important inscribed with gold or silver ink on opulent purple parchment. Hebrew also played an important role; Palermo's Jewish community participated in both the political and cultural life of the city. This multilingualism directly reflected Roger's determination that all the peoples of Sicily should feel included and protected, while also emphasizing and legitimizing the power of the monarchy.

George of Antioch, Roger's chief minister from 1126 onwards, encapsulated the cultural mix. A Greek who began his career with the Byzantine administration in Syria, he then moved to the Muslim court of al-Mahdiyya, in Tunisia. By the time he arrived in Sicily, he was fully acquainted with both the Byzantine and Arabic court traditions and systems of government—which he brought with him to his new role. George was instrumental in Roger's extensive building programme, making artistic suggestions and sourcing craftsmen and materials from the East and North Africa. He even built his own church, the iconic Concattedrale Santa Maria dell'Ammiraglio. This beautiful building exemplified the new hybrid Sicilian style; designed on a Greek-cross plan, with Islamic arches and niches combined with Norman arches, it was lavishly decorated inside with Byzantine mosaics, one of which depicted Roger being crowned by Jesus Christ himself. Roger had his own splendid Cappella Palatina constructed inside the imposing palace he "built with amazing effort and astonishing skill out of squared stones," in Palermo.[12] It is just as impressive now as it must have been in the twelfth century. Byzantine artists created the dazzling array of mosaics on the walls, Arab craftsmen designed intricate inlaid patterns in the marble floors, while the ceiling was constructed with hundreds of wooden panels, creating a labyrinthine, three-dimensional canopy, and covered in miniature paintings of daily life at court.

Roger transformed Palermo, creating a capital city fit for a king—which, in 1130, is exactly what he became. He had inherited the title of Count of Sicily from his father, but was raised to the dukedom of Apulia and Calabria when his cousin, William, unusually for a de Hauteville, died childless. Roger smoothly stepped into the breach, using his considerable military and naval forces to see off opposition from the Pope and various local nobles. In 1130, he united Southern Italy with Sicily, demanding the title of king as part of the peace settlement he negotiated with the Vatican. He was crowned in Palermo Cathedral on Christmas Day, and a monk who attended the ceremony later wrote, "When the Duke had been led to the archiepiscopal church in royal manner and had there through unction with the Holy Oil assumed royal dignity, one cannot write down nor indeed even imagine quite how glorious he was, how regal in his dignity, how splendid in his richly adorned apparel. For it seemed to the onlookers that all the riches and honour of this

22. The imposing facade of the Norman Palace in Palermo.

world were present."[13] The need to legitimize Roger's position radi-
ates from this account, unsurprisingly; in just a couple of centuries,
the de Hautevilles had managed to elevate themselves from heathen
raiders to anointed kings—a feat of quite astonishing magnitude
and audacity. Now, all they had to do was keep hold of their power.
They succeeded, but not for long; by the end of the century, the de
Hauteville name had died out and the crown of Sicily had passed to
the Hohenstaufens.

Roger II ruled as King of Sicily for twenty-four years, until
his death in 1154. During that time, he welcomed scholars to his
court in Palermo and led the elite in patronizing and encouraging
scholarship. The apogee of his intellectual legacy is a geographical treatise that he commissioned from one of his closest advisors,
the Arab scholar Abu Abdalluh Muhammad ibn Muhammad ibn
Abdullah ibn Idris al-Sharif al-Idrisi, who had arrived in Palermo
in 1138. Written in Arabic, its original flowery title, the "Entertain-
ment for He Who Longs to Travel the World," was later shortened
to the altogether less romantic "*Tabula Rogeriana*," or "The Book
of Roger." A groundbreakingly detailed description of the world,
encompassing rivers, mountains, climate, peoples, mercantile activity and distances between places, it was "the first serious attempt
to integrate three classical Mediterranean traditions of Greek, Latin
and Arabic scholarship in one compendium of the known world."[14]
In it, Idrisi took geographical knowledge from both the East and
the West, underpinning it with Ptolemy's cosmographical system of
the seven climes, or climate zones. In the preface, he explains that
Roger commissioned the work because he was interested in learning about the world, but also for more practical reasons: "he wanted
to know his lands in a wide-ranging and exacting way."[15] Ever the
pragmatist, Roger II was gathering information that would help him
to rule more effectively, but he died before the work was complete.
However, his heir, William I, was able to make use of Idrisi's book,

which was illustrated with seventy regional maps and a spectacular planisphere—a map of the world—made from pure silver. In the introduction to his translation of Plato's *Phaedo*, Henricus Aristippus claims that William was an incomparable king, "whose court is a school, whose retinue is a Gymnasium, whose own words are philosophical pronouncements, whose questions are unanswerable, whose solutions leave nothing to be discussed, and whose study leaves nothing untried."[16]

This was the world in which the anonymous scholar we met earlier found himself. Having finally got hold of *The Almagest*, presumably after persuading Henricus Aristippus to hand it over, he quickly realized that he had neither the necessary astronomical knowledge, nor the necessary Greek, to begin work. He threw himself into studying Euclid's *Data*, *Optica* and *Catoptrica*, along with Proclus' *De motu*.[17] At this point, he had a piece of luck: he met another member of the elite, Eugenius, an erudite, Greek-speaking Byzantine, who also knew Arabic and Latin. As a senior member of the royal household, Eugenius' duties were wide-ranging. They included the issuing of proclamations, overseeing accounts and setting boundaries. During his long career, he served several Sicilian monarchs and, in 1190, King Tancred promoted him to the rank of emir or admiral. Eugenius also managed to fit in a considerable amount of scholarly activity around his official duties, translating Ptolemy's *Optica* from Arabic to Latin, and two oriental texts from Greek into Latin. With Eugenius' help, the scholar was able to translate *The Almagest*, and it is thought that they finished work on the manuscript in the mid 1160s, pre-dating Gerard of Cremona's version by several years. It was a momentous event in the history of science—the first time that Ptolemy's great work could be read in Latin in its entirety—but, though pioneering, the Sicilian translation was not nearly as influential as Gerard's. Only four copies of it survive and, of these, only one is complete.

The anonymous translator of *The Almagest* wrote a detailed preface at the beginning of the work—our only source of information about him and how he came to translate it. Frustratingly, he does not tell us who he is, or where he comes from, but he was almost certainly not native to Southern Italy. We know even less about a translation from the Greek of Euclid's *Elements*, made in Sicily around the same time, but it was probably by the same man—there are many similarities in the translation style and vocabulary. Since, as we know, anyone studying *The Almagest* had to read *The Elements* first, it makes sense that this translator either began with *The Elements* or had already worked on it.

Consequently, there must have been a Greek manuscript copy of *The Elements* in Sicily in the mid eleventh century, which naturally leads to the question of where it had come from. The most likely source is Constantinople. We know that Henricus Aristippus had been given a copy of *The Almagest* there, and it doesn't seem unreasonable to suggest that the Byzantines also gave him a copy of *The Elements*. Additionally, the surviving copies of the Sicilian translation in Latin are similar to Arethas' Greek copy that was made in Constantinople, which is now in the Bodleian Library in Oxford. This has led some scholars to suggest that Henricus Aristippus got hold of this very book and took it to Sicily, where it was translated into Latin and from where it eventually made its way to England several centuries later. The web of transmission of these manuscripts is extremely complex, but it is possible to trace definite, yet delicate, connections. As we will see, this version of *The Elements* enjoyed more influence than the same translator's *Almagest*. It was the only translation from Greek made in the twelfth century, and it stands alongside the Arabic–Latin translations by Gerard of Cremona and Herman of Carinthia that we have already looked at. It is another translation from Arabic by Adelard of Bath, however, that dominates the transmission of Euclid's work in this period.

Adelard is a gloriously colourful figure in the history of medieval science. While Gerard of Cremona scribbled away diligently in the cathedral precincts of Toledo, the "man of Bath" swaggered around Southern Italy and the Middle East, consorting with kings, surviving earthquakes and generally living life to the full. There is no doubt that Gerard's contribution to scholarship was larger and more important, but, as a personality, he has faded into history. Adelard, on the other hand, has survived the last eight centuries remarkably well. He seems very much alive, in spite of the fact that all we have to build on are small fragments of information, faded inscriptions scribbled on manuscripts and oblique references in prefaces. He was obviously a gifted and somewhat eccentric individual, a talented musician (so talented, in fact, that he was asked to perform for the French queen), who enjoyed hawking as much as astronomy. Ambitious, adventurous and a bit of a show-off, Adelard was born in England, into the first generation after the Norman conquest, a time of great change and, for some, opportunity. He was lucky to be born into a wealthy family, who were connected to the powerful local bishop, Giso of Wells. He was educated in Bath, just as the diocesan centre was moved there from Wells by Giso's successor, John of Tours, who quickly began rebuilding and reviving the city.

Clever young Adelard must have benefited from this, but, having exhausted the educational opportunities on offer in his home town, he was then sent, probably on the recommendation of Bishop John, to the cathedral school in Tours, in the Loire Valley. It is possible he had already begun studying the sciences in England, and he certainly pursued them further in France. He would have been introduced to Euclid's *Elements* via the small fragments of Boethius' translation, which was then the basis of the mathematical curriculum.

Adelard was a talented scholar, but also a bit of a dandy who wore a bright green cloak and an emerald ring. In a manuscript copy of his *Regule abaci*, made in Paris in around 1400, there is a picture of him teaching Arabic numerals and duodecimal fractions. While it isn't a faithful portrait, it depicts him with long hair and a luxuriant beard, wearing an elegant red cape with a lapis-blue undershirt and a colourful striped hat. His writings reflect this dichotomy. Half are urbane, literary discourses designed to educate young noblemen, presented in elegant Latin, in the form of a dialogue between Adelard and his nephew—a literary device almost certainly borrowed from Plato—and he dedicated his work on the astrolabe, *De opere astrolapsus*, to his pupil Henry Plantagenet, the future King Henry II. *De opere* contains some scientific material and Adelard incorporated an introduction to the abacus into his *De eodem et diverso*, an allegorical discussion of the seven liberal arts. Similarly, the *Quaestiones naturales* includes some skilful examinations of the causes of natural phenomena. He had a talent for communicating complex scientific ideas and adapting them for an amateur, but interested, audience. These three works enhanced Adelard's professional position and would have made him money, allowing him to spend time on his serious academic interests, which constitute the other half of his writings. The most important of these are his translations from the Arabic of al-Khwarizmi's *Zij*, Abu Mashar's *Introduction to Astrology* and *The Elements*. These works are concise and scientific, informative rather than flowery—and they are not dedicated to anyone. Adelard wrote them for himself and his students, for serious study. And, as ever, the big question is, where did he find them?

There are various possibilities. Adelard travelled widely, taking books and ideas with him from place to place, meeting scholars and connecting them to a wider network. He provides a definitive link between the great centres of medieval scholarship. Early in the twelfth century, Adelard set off on his great journey, parting company with his nephew and some other pupils, near Laon, in France. He had become restless and disillusioned with the intellectual life in Northern Europe. The arguments and ideas preoccupying scholars in France seemed pointless; Adelard's outspoken view was that they were merely "making ropes out of intellectual sand."[18] The world was opening up, merchants were arriving with thrilling tales and exotic goods, the Normans had conquered Sicily and Southern Italy, the First Crusade was under way. Adelard—curious, adventurous, courageous—couldn't resist. As he later explained in his treatise *Quaestiones naturales*, he travelled south, determined to broaden his horizons and his knowledge, "among the Arabs."[19] As far as we know, he spent the next seven years on the road, visiting Rome, Salerno, Sicily, Greece and Asia Minor.

We do not know exactly where Adelard went, but he must have taken the major land route from Northern Europe to Rome, the Via Francigena. Popular with pilgrims, it passed through Laon, then on to Rheims. From there, it led south into what is now Switzerland, through the Alps at the San Bernardino Pass, where enterprising locals had set up toll gates at which they charged travellers to cross. (This was a lucrative source of income, as long as the Normans weren't passing through. They smashed the barriers, stabbed the toll collectors and marched on into Italy—the normal rules simply didn't apply.) Once over the Alps, the road would have led Adelard down onto the great plains of northern Italy and through the busy markets at Pavia. From there, it was down towards the coast, passing through Lucca, Siena, Viterbo and, finally, Rome. Adelard could easily have joined a band of pilgrims or merchants in Laon and travelled south with them. As we saw in the previous chapter, the road from Rome to Salerno was well established, and would have taken Adelard past Montecassino. Indeed, he might well have spent the night there, particularly if he was travelling with a party of pilgrims. Most large monasteries had accommodation for travellers, especially people on any kind of religious mission; they provided simple food and basic facilities for a small fee. Since Adelard did not, however, use any of Constantine's translations, there is no firm intellectual link. But he was definitely influenced by Salernitan texts and Galen's theory of the four humours, and used them in his *Quaestiones naturales*. It isn't difficult to imagine Adelard's excitement at being in Salerno, studying with the city's renowned doctors. The *Quaestiones* also drew on Alfano's translation of Nemesius' *De natura hominis*, which would also have been available in Salerno, making Adelard a definite channel of transmission from Southern Italy to Northern Europe. He describes leaving Salerno and meeting a Greek, with whom he discussed medicine and other scientific questions, such as magnetism. Even though he would only write about this event years later, his delight in encountering knowledgeable people who were interested in science shines through.

The available dates hint that Adelard's next stop was Sicily, where he marvelled at Mount Etna and probably spent time in Syracuse. He dedicated a treatise to the ancient city's bishop, a man named William, who was "most learned in all the mathematical arts."[20] This suggests that they discussed mathematics, and it is at least possible that William either gave Adelard a copy of *The Elements*, or encouraged him to look for one on his travels in Antioch and Asia Minor. He certainly had an Arabic copy after he had returned to England, because he used it as the basis of his translation.

The First Crusade had opened up routes between Southern Italy and the Eastern Mediterranean coast, and when Adelard was in Antioch, the city was ruled by Robert Guiscard's grandson, Tancred, so it had close connections with Sicily. This made the area more accessible, but it was also turbulent and dangerous. The Crusader states were constantly at war with one another, and with the Turks; travellers like Adelard had to be careful. Occasionally, the violence had a positive side effect. In 1109, the *Damascus Chronicle* records, "The Franks pressed their attack upon the city [Tripoli] . . . the quantities of merchandise and storehouses and the books of its college and in the libraries of private owners, exceed all computation . . ."[21] Many of these books fell into the hands of the Genoese who led the attack; some, at least, must have ended up either on sale or being copied by scholars, before being shipped to Italy and on to France, Germany and England.

Adelard returned to England in around 1116. He was made an official in the government of Henry I, but he also began to translate the texts he had encountered on his travels, including his influential Latin translation of Euclid's *Elements* from Arabic. Confusingly, there are three different versions of this text, all initially attributed to Adelard, and therefore known as Adelard I, II and III. I is the basic translation, II is based on I and on various other translations, while III is a commentary on version II. Recent investigations have shown that Adelard II was probably made by Robert of Chester, in Spain, and is the version included by Thierry of Chartres in his *Heptateuchon*, making it the most influential of the three. In the thirteenth century, both II and III were used by the Italian scholar Campanus of Novara to create an alternative edition that went on to be the basis of the first printed edition. Untangling the web of connections between the different versions of a text is extremely difficult and, at times, intensely confusing. But it is fascinating to see how frequently manuscripts moved from place to place, and that scholars managed to get hold of a variety of different copies in order to produce their own versions. There was clearly a well-connected community of intellectuals able to communicate with one another across huge distances. As we have already seen, the network of Benedictine churches and monasteries underpinned this web of interaction, connecting northern Spain with France, England, Germany and Italy, as clerics travelled between them with news, correspondence and, of course, books.

When Adelard returned to England, he translated al-Khwarizmi's *Zij* (astronomical tables)—specifically the version adapted to the coordinates of Córdoba by Maslama al-Majriti. This has raised the possibility that he also travelled to Spain at some point, but there is no evidence to support this idea. In which case, Adelard must have got hold of this book from someone who had been in Spain, or had contacts there. The most likely candidate is Petrus Alfonsi, an intriguing figure who was born into the Jewish faith in Huesca, in northern Spain, but converted to Christianity in 1106, with Alfonso I of Aragon acting as his godfather. He later wrote a refutation of the Jewish faith in the form of a dialogue between his two selves, before and after his conversion. Alfonsi benefited from an excellent education in languages—he was fluent in Arabic, Hebrew and Romance, must have known Latin, and no doubt learned English during the years he spent in England. Huesca was also within the sphere of influence of Zaragoza, with its distinguished tradition of mathematical and scientific study, which also benefited Alfonsi. One source suggests that he worked for King Henry I of England, as his physician, and he was certainly part of a circle of astronomers and philosophers in the West Country—indeed, it may well be because of his encouragement and the books he must have brought with him to England that the study of astronomy began to flourish in the country at that time. It is also likely that he met Adelard when in the vicinity of Bath, and that they began working together on translations of the *Zij* and *The Elements* soon afterwards.

Adelard's works, both his own and the translations, were particularly well dispersed. The anonymous scholar in Sicily quoted a phrase from the *Quaestiones naturales* in the preface to his *Almagest*, suggesting he had, at the very least, read it. Some of the glosses to Adelard's *Elements* contain little notes that give a revealing glimpse of his circle: "Only Adelard will be able to understand this problem," says one aside; "this proposition of ours is proved true without any help from John," quips another; and, best of all, "Goodbye, Reginerus. Whoever does not know how to reply to you should present you with a white cow!"[22] We can't assume that all the scholars we have met worked in a similarly humorous and cooperative atmosphere. Adelard does seem to have been unusually witty and cultured, and it is thrilling to hear the voices in his study, echoing back at us through time, vivid and alive. These comments show what a collaborative project translation could be, some of the first hard evidence for something we have suspected at every stage of our journey through the Middle Ages. Here, at last, we are given a glimpse into one of the "Houses of Wisdom," and can, albeit briefly, hear the chatter of the translation room.

The question of whether Adelard actually learned Arabic himself is a tricky one. He could easily have picked up a degree of spoken Arabic on his travels, but learning to read and write the language was another matter. Given the political situation, it would have been hard for him to make contact with Muslim scholars during his travels in the East. It seems that Adelard was fascinated and impressed by Arabic science, but was not as knowledgeable about it as he liked to appear; modern scholars point to the lack of specific Arabic sources in his writings and have concluded that he obtained his knowledge of Arabic science and language orally, rather than by reading. For example, when Adelard travelled to Tarsus, in Cilicia, he learned human anatomy from an old man, who demonstrated how sinews function by suspending a corpse in flowing water. He also describes hiding under a bridge near Antioch when an earthquake struck, a detail that provides us with the date: 1114. Antioch, founded in the fourth century BC by one of Alexander the Great's generals, is on the banks of the Orontes River. With Seleucia Piera, its seaport on the Mediterranean, Antioch was a major hub on the Silk Road, growing so prosperous that at times it even rivalled Alexandria. Communities of Jews had lived there for centuries, as had some of the earliest Christians. By the time the crusaders took the city, in 1098, it had been ruled by the Arab Empire, the Byzantines and, briefly, the Seljuk Turks. This made for a potent cultural mix and plenty of opportunity for the exchange of ideas. Arabic was Antioch's main language, and the city soon became an important base for Western Europeans in the Middle East. The Pisans were the first to take advantage, closely followed by the Venetians and the Genoese. The Pisans had supported the crusaders with ships and naval backup. Their reward was a quarter of the city of Antioch, where they settled in 1108, founding trading posts along the coast and organizing merchant fleets, which carried huge cargoes of spices, sugar, cotton, wine and expensive fabrics back to Italy. There were Pisan and Venetian quarters in Constantinople as well, part of the growing network of European influence in the Levant and Eastern Mediterranean.

Throughout this story we have seen how trade opens up routes through which knowledge and ideas flow, propelled by merchants and diplomats, who were often scholars, too. Several Pisans fell into this category. In the early twelfth century, Stephen of Antioch left Pisa and travelled to Syria, where he learned Arabic and produced a new translation of Ali ibn al-Abbas' *Kitab Kamil*, because he considered Constantine Africanus' version, the *Pantegni*, not only inadequate but inaccurate. Stephen might have been in Antioch at the same time as Adelard, and, if so, it is tempting to speculate that they met. The scholarly community would have been extremely small, so it is not beyond the realms of possibility. Adelard would certainly have sought out like-minded individuals wherever he went. Stephen could even have inspired Adelard's later translating activity and helped him to obtain manuscripts.

There is one piece of evidence that suggests Adelard acquired books in Antioch—one that takes us briefly back to Spain. John of Seville wrote in the preface to his translation of Thabit ibn Qurra's *Book of Talismans* that the copy he worked from had been owned by a man from Antioch. Today's foremost scholar of this period, Professor Charles Burnett, has suggested that this man could have been Adelard, who translated the *Book of Talismans* himself after his return to England, comically exchanging Baghdad for Bath in the spell for ridding a city of scorpions.[23] Having completed his travels in the East, Adelard would have had no problem boarding a merchant vessel—in Tyre, or the port of Antioch itself—that was heading west to Italy or Sicily, from where he could make his way slowly back to England.

Adelard's visit to Sicily took place at a time when Latin culture was beginning to dominate; the island's Muslim population had either converted to Christianity or emigrated to al-Ándalus or North Africa. Roger II's heirs, William I ("the Bad") and William II ("the Good"), were less interested in Arabic culture and also, crucially, less skilful at keeping control; the resulting unrest encouraged wealthy Muslims, for example Roger II's confidant al-Idrisi, to leave Sicily in search of somewhere more conducive to a peaceful and productive life. But, although the Muslim population declined, the island was still an important stopping-off point for all manner of people travelling across the Mediterranean. In 1184, an Andalusian pilgrim called Ibn Jubayr arrived there on his way home from Mecca. He stayed in Sicily for the month of December and left us with detailed descriptions of what he saw. He was impressed by the landscape, paying the island the ultimate compliment of being "a daughter of Spain in the extent of its cultivation," and he said of King William II, "He pays much attention to his (Muslim) physicians and astrologers, and also takes great care of them. He will even, when told that a physician or astrologer is passing through his land, order his detainment, and then provide him with means of living so that he will forget his native land."[24] Ibn Jubayr even claimed that William could read and write Arabic, showing that, despite the new dominance of Latin, Sicily was still a polyglot culture.

Towards the end of the twelfth century, William II died, leaving no heirs. The Sicilian crown passed to his aunt, Constance, who

23. An illustrative scheme of mosaics in the Martorana Church (wall panel) and Palatine Chapel (columns).

was married to the Hohenstaufen Holy Roman Emperor, Henry VI. The Sicilians were not overjoyed at the prospect of being ruled by a German dynasty, nor at being subsumed into the Holy Roman Empire. Four years of fighting ensued before Henry VI was able to assert control and have himself crowned in Palermo Cathedral, on Christmas Day 1194—a nod to the memory of his wife's father, Roger II. Constance was not at his side; she was on the mainland, near Ancona, giving birth to their only child, Frederick. Just three years later, Henry was dead. Constance wisely had her little boy crowned to help ensure his succession, but she died the following year, leaving him an orphan. According to legend, Frederick grew up on the streets of Palermo, learning to speak six languages and being cared for by his citizens. When he came of age, in 1208, he immediately began taking back control from the noblemen who had seized power during his minority. The young emperor was a mythical figure from an early age, so brilliant, so talented, so dazzling that he was known simply as *"Stupor Mundi"*—Wonder of the World.* The de Hauteville name may have died out, but Roger II's spirit of intellectual curiosity survived and flourished in his grandson Frederick, who patronized scholarship and encouraged translation. His Hohenstaufen grandfather, Frederick Barbarossa, had also been an enthusiastic promoter of learning, granting privileges, in 1158, "to all scholars who are travelling abroad in the cause of learning . . . through whose learning the world will be enlightened and the life of citizens enriched."[25] Frederick II continued and extended these protections, galvanizing the world of scholarship and enlarging the universities. This continued so that, "by the end of the Middle Ages

---

* Other nicknames were not so flattering; the papacy, who excommunicated him no less than four times, called him "the Antichrist" and "the Punisher of the World."

thousands of students were on the road"—something that caused a huge expansion in the dissemination of ideas.[26]

The imperial court was also constantly on the move. Even though Palermo remained the capital of Frederick's empire, he hardly spent any time there as an adult. His vast dominions claimed his attention, and, when he was in the south, he was happier in Apulia. He also favoured Naples, and it was the university he founded there that eclipsed the medical school at Salerno. Frederick's court attracted the most talented and ambitious men of the age. Of the many scholars who orbited around him, two stand out: Michael Scot and Leonard of Pisa, known as Fibonacci.

Like Adelard a century before, Scot travelled extensively, leaving his home in Scotland to pursue his education, and it is thought that he studied at Durham, then Oxford and Paris. Inevitably, there are many gaps in his itinerary, but he was certainly in Toledo on 18 August 1217, as it was on that day that he signed and dated his translation of al-Bitruji's *On the Sphere*. This text, and Scot's translations of Aristotle's *De animalibus* (*On Animals*) and a commentary by Averroes on Aristotle's major work on cosmography, *De caelo et mundo* (*On the Heavens and the Earth*), were all produced in that city, and copies remained there. Scot then travelled to Italy, taking copies of his work with him, and they were soon circulating there, too. By this time, Scot had mastered Arabic, which suggests that he was in Spain for some time, learning the language and studying with local Mozarabs. He must have had considerable proficiency in maths and science in order to produce these translations, for which he also drew on many of the texts that had been translated into Latin a few decades earlier by Gerard of Cremona and the other Toledan translators. Scot was closely connected to Rodrigo, Archbishop of Toledo, and would have known the younger members of Gerard's circle— men like the great translator of Galen, Mark of Toledo. Michael

Scot is therefore a major channel for the movement of books and ideas from Spain to Italy in the early thirteenth century, as well as an important member of the generation of translators in Toledo who succeeded Gerard and his colleagues.

According to Pope Gregory IX, Scot knew Hebrew as well, which he could also have picked up in Toledo, from the community of Jewish scholars there. Later in life, he also corresponded with a Jewish scholar in Palermo. Many strange and wonderful stories have evolved about Michael Scot, for the most part based on his fame as an astrologer—a profession that teetered on the edge of acceptability, esteemed and depended on by secular rulers, but periodically denounced and demonized by the Church. Legend has it that he predicted his own death from a blow to the head by a falling stone. In order to avoid this fate, he constructed a metal helmet, which he wore at all times, but unfortunately it did not protect him when a piece of masonry fell from the ceiling of a church when he was attending mass, killing him instantly. Michael's interests were wide-ranging, as you would expect of a gifted scholar in this period. He was a physician, well acquainted with both Salernitan medical teaching and Arabic doctors like al-Razi. He used these sources in his most popular work, *Physionomia*, which deals with generation and prognostication, using dreams and human anatomy. He drew on *The Almagest* and the *Toledan Tables* in one of his works on astronomy, but it is his translations of Aristotle—which he made available in Latin for the first time—that are his most significant textual legacy.

Scot arrived in the rapidly growing university city of Bologna around 1220. We can be absolutely sure that his luggage was filled with books, but, apart from that, we have no idea how he travelled. Frederick II was in Bologna at the same time, and it's quite possible that this is when they first met, and Scot entered his service. He quickly rose to prominence and spent the rest of his life with the imperial court, travelling wherever it went. He and the Emperor grew close; they shared many interests and discussed them at length. They also carried out experiments together, testing the effect of bloodletting when the moon was in Gemini and attempting to measure the heavens using a tower. Frederick overruled the Church to allow human autopsies to be performed for the first time since AD 150, while Scot produced detailed case studies on his patients. There are echoes, here, albeit on a more modest scale, of the House of Wisdom in Baghdad. Like the Caliph al-Ma'mun, Frederick II asked questions, famously addressing a list of enquiries to scholars and rulers across the empire and the Islamic world. The answers he received were recorded by the Arab philosopher Ibn Sab'in in *The Sicilian Questions*. They are a window into the medieval mind, revealing the intellectual preoccupations of the time and the limits of knowledge. Many of the questions stem from close observation of the natural world, such as, "Why does an oar, a lance or any straight object partly submerged in clear water, appear curved (or rather: bent) towards the surface?" and, "The Emperor asks why the star Suhayl (Canopus, in the constellation Carina) appears larger to the eye when rising than at its perigee?" Some Muslim writers seemed to think such questions were posed to test them, but Frederick did not know the answers; they were genuine attempts to spark intellectual dialogue. Scot himself provided some responses in one of his own works, explaining that the earth is round, like a ball, but surrounded with water, like a yolk in an egg, moving on to discuss volcanic activity, a phenomenon which had preoccupied Sicilian rulers, visitors and scholars for centuries.

There was no better place in Europe to be a scholar than Frederick's court. It was the centre of intellectual brilliance, but one that moved constantly: from Palermo to Naples, Bologna to Pisa, Brescia to Padua, Vienna to Verona, Frankfurt to Konstanz, Brindisi to Jerusalem—an itinerary that would make even the most seasoned traveller's head spin. For the first time in this story, the centre of academic life was not fixed in one location, profoundly increasing the opportunities for the transmission of knowledge and providing a ready-made network through which it could flow. Scholars of all faiths and cultures were welcome, so long as they could keep up. The most brilliant of them all was Leonard of Pisa, known as Fibonacci. The product of the Pisan trading empire, he was educated by the finest Arab mathematicians in Bougie (modern-day Béjaïa), on the coast of North Africa, where his father worked for the Pisan Chamber of Commerce. This enabled him to combine the theoretical genius of al-Khwarizmi's algebra, the Hindu-Arabic numerals and system of positional notation with the practical demands of Pisan commerce. Like Michael Scot, this exceptionally talented young man found his way to Frederick's court, where he produced books that became the foundation of the study of mathematics in Western Europe. His *Liber de numero* (also known as the *Liber abaci*), originally written in 1202, set out the principles of arithmetic and helped to popularize the Hindu-Arabic numeral system in Europe.

In 1227 or 1228, Fibonacci produced a new edition of this book, dedicating it to Michael Scot, who had requested a copy. In the preface, Fibonacci also mentions that he has written a book called *Practica geometriae*. Building on *The Elements*, this treatise especially focuses on the irrationalities listed in Book 10, using algebra to devastating effect in differentiating between the roots of cubic equations and quadratic irrationalities. This was apparently a task he was set by another scholar at Frederick's court, John of Palermo, and the two men discussed these mathematical problems with the Emperor himself while the court was in Pisa.

The dialogue between Christian and Muslim scholars in this period shows how they moved between their two worlds, setting each other challenges, working together, sharing ideas and pushing the boundaries of knowledge. Frederick's interest in the Arab world, established during his childhood in Palermo, only increased when he travelled to the Holy Land in 1228/1229 and experienced it for himself. He marvelled at the luxury of the lifestyle, tried innovations in falconry, learned how to play chess and admired the emphasis on learning at the sultan's court; when he returned to Europe, he took many of these novelties with him.

Sicily held on to its position as a hub of Mediterranean trade, but, in other respects, her star was on the wane. As he strove to rule over his vast empire, the *Stupor Mundi* left his childhood home behind. Because cultural activity in Palermo was, for the most part, concentrated within the walls of the palace and the dominions of the court, it declined significantly when the court moved on. Frederick's establishment of a university in Naples ensured that learning continued in Southern Italy, but at the expense of both Palermo and Salerno. In the following century, Naples' position was further enhanced when it became the home of the new King of Sicily, Charles of Anjou. Charles kept his predecessors' traditions of scholarship alive, albeit in a modest way. He encouraged the scholar Niccolò da Reggio to translate a large number of Galen's works from Greek into Latin, continuing the Sicilian tradition of bypassing the Arabic versions if they could find copies in the original Greek. This aspect of scholarship prefigured the humanist movement by several centuries; as we shall see in the next chapter, the obsession with returning to the original Greek source material would become a defining feature of the Renaissance intellectual world, to the detriment of the contribution of Arabic scholarship.

The de Hauteville name had long since passed into memory, but the Normans' extraordinary journey from cattle raiders to kings is one of the great tales of the medieval period. Their sphere of influence, stretching from the far north of England to the shores of Southern Italy, and beyond, into the Middle East and the heart of Jerusalem itself, opened up lines of communication that enabled the exchange of ideas on a hitherto unprecedented scale. The wandering scholars, who set off into the unknown in search of wisdom and enlightenment, were key agents in the transmission and transformation of knowledge in this new, connected world, learning, teaching and spreading ideas. The Normans made Sicily into a major power, a hub at the centre of the Mediterranean world, where ideas moved between cultures. In their dazzling court in Palermo, they brought scholarship in Europe into the secular sphere for the first time since Christianity had taken hold, creating a blueprint that was copied in the courts of Europe for centuries. The traditions of the Byzantine and Muslim empires arrived in Europe through their agency, profoundly changing court culture and the ways in which power was expressed. The Normans stand beside the Umayyad and Abbasid caliphs in the pantheon of rulers whose personal intellectual interests and talents expanded the frontiers of science.

By the time Frederick II died, in 1250, the world was changing. The great Italian mercantile powers now began to define the geopolitics of the Mediterranean, and, in northern Italy, the growth of independent city states heralded a new era—the Renaissance.

EIGHT

Venice

A place more like an entire world than a city.

—Aldus Manutius

I was conducted through the longest street, which they call the Grand Canal, and it is so wide that galleys frequently cross one another; indeed I have seen vessels of four-hundred tons or more lie at anchor just by the houses. It is the fairest and best-built street, I think, in the world and goes right through the city. The houses are very large and lofty, and built of stone; the old ones are all painted; those of about a hundred years standing are faced with white marble from Istria, about one-hundred miles off, and inlaid with porphyre and serpentine. Within they have, most of them, at least two chambers with gilt ceilings, rich chimney pieces, bedsteads of gold colour, their portals of the same, and exceedingly well furnished. In short, it is the most glorious city that I have ever seen, the most respectful to all ambassadors and strangers, governed by the greatest wisdom, and serving God with the most solemnity.

—French ambassador Philip de Comines
(*c.*1447–1511), touring Venice in 1495

THE STORY OF VENICE begins in the fifth and sixth centuries, as the Roman world crumbled from within and was attacked from without. The hard, straight roads that had carried Roman legions, merchants and pilgrims efficiently around the empire for centuries became avenues of terror as invading armies marched down them towards Rome. On their way, they passed the great cities of northern Italy—Aquileia, Altino and Padua, pausing only to wreak havoc by siege, sword and flame. Those who managed to escape fled towards the sea, carrying the few possessions they had been able to save. When they reached the water's edge, they found themselves in a strange new world. In the north-eastern corner of Italy, there is no clear definition between land and sea, no cliffs with bays and beaches, no rocky division between the two realms. Here, where the coast curves around the top of the Adriatic Sea, the two elements unite across a vast, flat expanse. The water slips over the fluctuating sands, islands appear and disappear, forests of reeds grow in the marshy ground, and the light, shining through billions of droplets of evaporating water, appears pearlized, supernatural, conjuring mirages on the horizon, a luminescent haze separating the bright blue of the sky from the pale aquamarine of the water.

Over millennia, the great rivers, the Po and Piave, had deposited huge quantities of silt into the bay, carried down from the mountains. The currents formed the silt into a curved line of sand bars, running parallel to the coast, creating a huge lagoon of shallow water in between, cut off from the open sea apart from a few channels that fed water in and out as the tide rose and fell each day. A haven for birds, fish and mosquitoes, but also for the refugees who had managed to reach the shifting, grassy islands in small, flat-bottomed boats—the only craft that could navigate the unpredictable waters. These tenacious people built their lives in this flat, watery world, protected by the sea that separated them from the mainland, but at the same time constantly threatened by the high tides, the *acqua alta*, that could inundate their homes at any moment, and occasionally floods the city to this day. Known as the Veneti, they learned to survive and, eventually, to flourish. They lived off the abundant fish in the lagoon and they sank huge tree trunks into the water as foundations for their houses, returning to the ruined cities on the mainland for stone, marble, bricks and wood—any building materials they could find and transport.

Small communities began to grow on the cluster of little islands in the centre of the lagoon. A society developed with its own particular system of government, ruled by a *dux* (Latin for leader, which, over time, morphed into the word *doge*), who was elected for the first time in AD 697 to rule over the nascent city of Venice. The Venetians were resourceful and determined. They laid bridges across the narrow channels of water, they constructed dams against high tides and drained the land, they built narrow, flat-bottomed boats that could dip and glide smoothly across the waters. They developed effective ways of making the most of life in the lagoon. The sea could not yield crops, but they made it profitable by constructing salt pans— areas of very shallow water that evaporated in the sun, leaving acres of shining minerals, which they broke up with rollers and rowed to the mainland to barter for wheat and barley. This lack of self-sufficiency forced them to trade, to sail not only up the great rivers to the markets at Cremona, Pavia and Verona, but also out into the open sea and down the Istrian coast. Controlling the Adriatic was fundamental to the Venetians' ability to trade in the Mediterranean and the East, and they soon established a string of trading posts along the coast, offering the inhabitants protection from the vicious pirates who terrorized the region, in return for power. In 998, the Doge of Venice added Dux of Dalmatia to his list of titles.

Right from the very beginning, the Venetians were independent. They turned their isolation to an advantage by keeping out of politics on the mainland, while focusing on trade and diplomacy. Geographically, their growing city was perfectly placed between the two major political powers of the time: the Byzantine Empire to the east, and the Frankish kingdom to the west. In 814, the inhabitants of Venice made a treaty that articulated their unique position. They would be a province of the Byzantine Empire, but would, at the same time, pay tribute to the Franks. This could have given them the worst of both worlds, but, in fact, it put the Venetians in a privileged space between the two empires and, most importantly of all, it gave them trading rights and the freedom to use Italian ports. In 1082, the Byzantines extended Venetian trading rights, exempting them from taxes and customs duties across the empire, marking another crucial moment for the city's commercial growth. By 1099, there was a lucrative spice trade with Egypt, and Venice was on course to create the most successful maritime empire the world had ever seen.

Stable, relatively democratic government, rigorous organization and an absolute devotion to the city lay at the heart of Venice's extraordinary success. That devotion was not only practical, it was religious, too. Venetians believed their city had divine foundations, they worshipped it, creating unusually high levels of loyalty and social cohesion. While the rest of Europe was yoked under the feudal system, with noble families tearing themselves and everyone around them apart in violent power struggles, Venice prospered as the first republic of the post-classical world. Its inhabitants were fervently united around a shared enterprise: the glorification of their beloved city, which they called *La Serenissima*—The Most Serene

Republic. This unity was born out of the challenges of living in the lagoon. The Venetians were forced to work together just to survive, to overcome the problems posed by their inconstant environment. This precarious existence meant that they prized stability above all else, especially when it came to the city's governance. Organization, cooperation and control were of fundamental importance to everyone's survival, and an efficient administrative framework soon evolved, overseen by the doge and patricians—members of the founding families of the city.

The city grew, but not in the same haphazard, sprawling way as cities on the mainland. Every new row of houses, every canal, every campo had to be carefully planned. Like Baghdad and Córdoba, Venice zoned different types of manufacture in different areas, its island structure perfectly suited to this form of town planning, a novelty in Europe at the time. This idea was probably brought back to Venice by merchants who had visited those cities and been impressed by their design and organization. The island of Murano became the centre of glass-making when the foundries were moved there, in the thirteenth century, to protect the city from fire—the roaring furnaces that smelted the glass posed a danger to its tightly packed, wooden buildings. From the twelfth century onwards, the north-eastern corner of the city was home to the Arsenale (from the Arabic *dar sina'a*, meaning "place of construction"), the Venetian shipyard, where a community of workers known as *arsenalotti*, numbering somewhere between 6,000 and 16,000 men, built ships of every kind, which were sold and sailed around the globe. This was the engine room of the Venetian Empire, the birthplace of its navy, its fleet of trading vessels and the warships that were eagerly purchased by major powers throughout the medieval and Renaissance periods. The Arsenale's greatest challenge came in 1204, when the Venetian state agreed to fit out the entire Fourth Crusade—a massive financial risk, but one that eventually turned out well. The Venetians regained control of the city of Zara, now Zadar, and were paid in full by the leaders of the Crusade. They even managed to orchestrate the redirection of the Crusade against Constantinople itself, and the resulting sack of the city, led by the legendary blind doge, Enrico Dandolo, furnished Venice with a vast sum of money and piles of priceless artefacts, including the four bronze horses which are now reproduced on the facade of the Basilica di San Marco—the originals are kept inside to protect them from the weather.

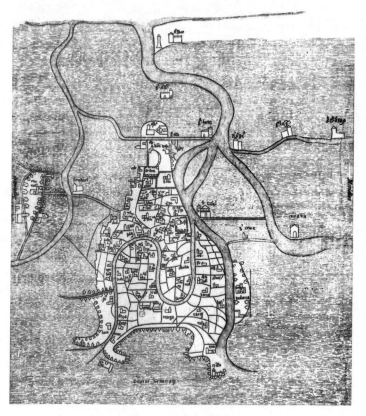

24. A map of Venice in the twelfth century, east at the top.

Founded by exiles, it is perhaps unsurprising that Venice has a history of welcoming strangers, and it became a destination for pilgrims and tourists alike from very early on. Enterprising locals opened taverns, like the Lobster, Hotel Luna and the Little Horse, and offered their services as guides in the Piazza San Marco, where stallholders sold snacks and souvenirs, just as they do today. These days, tourism is the biggest (and, in many ways, the only) industry in the city. Around thirty million people come each year, to a city with only 54,000 inhabitants, earning Venice a reputation as the Disneyland of Italy. Visitors come to gaze in wonder at the floating city, with its watery thoroughfares and heartbreaking beauty. Standing in the Accademia Gallery, in front of a Renaissance painting of the city, it is striking how little it has changed. The architecture, the bridges, the gondolas all look the same; apart from the elaborate clothing, the only noticeable difference is that satellite dishes have replaced the many chimneys on the rooftops. Venice seems frozen in time, a historical theme park where the modern world does not really intrude, a place where beauty and antiquity prevail, where even dilapidation is imbued with splendour. Many modern tourists are just day trippers, arriving on the giant cruise ships that are controversially allowed to dock right in the lagoon. In the Middle Ages, visitors stayed much longer, often settling and making Venice their home for several years, encouraged by the city's atmosphere of acceptance and enterprise. In the twelfth century, a steady stream of German merchants began to arrive, establishing themselves in the busy area near the Rialto Bridge, where, in 1228, they built their headquarters, the Fondaco dei Tedeschi. These adventurous northerners were part of a tide of immigrants swelling the population of Venice, which, by 1300, had reached 120,000.[1]

The other significant community of foreigners was that of the Greeks, who came to live and trade in Venice in large numbers. They brought their ancient culture and language with them, and might well have been one of the reasons that the poet and scholar Petrarch came to Venice in 1351. He wanted to learn Greek so that he could translate the classical texts he had collected on his travels, which would provide the foundation of the movement that became known as humanism. Petrarch was a friend of the doge, Andrea Dandolo (1306–1354), himself credited with helping to launch the Renaissance in Venice with his lucid, well-researched history of the city, and the two men enjoyed a lively intellectual relationship. After Petrarch left the city, this continued by letter, a correspondence that was carried on after Dandolo's death by his secretaries, so keen were they to keep abreast of scholarly developments in the rest of Italy. They even persuaded Petrarch to return to Venice, hoping that he would leave his exceptional collection of books to the city when he died. Unfortunately, the plan was ruined by an argument about Aristotelian logic with some of Venice's patricians. Furious, Petrarch loaded his manuscripts onto a boat and sailed back to the mainland, never to return.

Petrarch was a founding father of the Renaissance in Italy, inspiring the next generation to collect manuscripts and promote learning by whatever means possible. He corresponded with and encouraged a young scholar called Coluccio Salutati (1331–1411), who went on to spend much of what he earned as Chancellor of Florence on a collection of 800 manuscripts—one of which was the Latin translation of *The Almagest* we encountered in Sicily. It was Salutati who established Florence as the centre of intellectual life in fourteenth-century Italy, and the major marketplace for classical texts. In 1396, he invited the Byzantine diplomat Manuel Chrysoloras to come and teach Greek in the city—the first time it had been possible to study the language as an academic subject in over a millennium.[*] Salutati

---

[*] Although, as we saw in the previous chapter, vernacular Greek had been spoken consistently in Sicily and parts of Southern Italy throughout the Middle Ages.

also had the foresight to write and ask Chrysoloras to bring as many Greek manuscripts from Constantinople as he could. Chrysoloras spent only three years in Florence, but his teaching—which, judging by the eulogies he received, was excellent—had a transformative effect not only on his own students, but also on future generations who used his textbook of Greek grammar. Dictionaries, grammars and other language aids were a vital aspect of the spread of learning in this period, a huge advantage to anyone trying to learn a new tongue or translate a manuscript. Previously, only a teacher or interpreter (as had been prevalent in Toledo and Sicily) could provide this kind of knowledge. The second part of Chrysoloras' course focused on translation from Greek to Latin, and eschewed the word-for-word style adopted by earlier translators like Gerard of Cremona, instead favouring an emphasis on the meaning of the text.

Chrysoloras' students were prolific translators, using their recently acquired language skills to produce new, improved editions of classical texts, translated directly from the original Greek. At the same time, intrepid scholars were setting off for remote monasteries in the mountains of Italy in the hope of finding ancient, overlooked texts that had survived the centuries in their libraries. Some, like Poggio Bracciolini, a scribe in the papal Curia, went even further afield, heading north over the Alps to Germany and Switzerland. In his intriguing book *The Swerve*, the historian Stephen Greenblatt describes expeditions made by Poggio, "the champion manuscript-hunter," to monasteries such as Saint Gall and Cluny Abbey. Poggio was also a champion letter-writer; his personality resonates through the missives he exchanged with his many friends and acquaintances. He was an erudite scholar, who worked for several popes and could quote Cicero to his heart's content, but also a man who wrote a collection of filthy short stories and was tickled by a visit to the public baths while travelling in Germany: "It is comical to see decrepit old women as well as younger ones going naked into the water before the eyes of men and displaying their private parts and their buttocks to the onlookers,"[2] he wrote to his friend, Niccolò Niccoli.

Poggio's explorations of monastic libraries in the mountains around Lake Constance yielded some interesting results; he found a few unknown speeches by Cicero and a manuscript containing the complete works of Quintilian, both of which were well received by his fellow humanists back in Italy. His final discovery, however, eclipsed all the others. In January 1417, on a dusty shelf in the depths of a monastery library, Poggio found a copy of *De rerum natura*, by the Roman philosopher Lucretius—a book that had not seen the light of day for centuries, about which there had only been rumours. Condemned and suppressed by the Church for over a millennium, this lyrical, complex epic poem contained ideas that were so icono-clastic, so threatening to the existing order, it was a miracle that it had survived at all.*

Lucretius (99–55 BC) was an Epicurean, a follower of a school of philosophy founded in Greece in the third century BC, based on the (as it turned out, visionary) belief that everything in existence is made up of tiny building blocks. The Epicureans named these blocks "atoms"—something that is so small it cannot be divided fur-ther, that cannot be created or destroyed. In the *De rerum natura*, Lucretius passionately espouses this and the host of ideas that fol-low on from it: there is no creator or divine scheme; everything in creation has evolved and continues to evolve, adapt and reproduce; human beings are just one of the millions of organisms on the planet with no central or unique role in the universe; there is no need to fear death, because the soul dies and there is no afterlife. These ideas are still highly contentious in some parts of the world today, so just

---

* The fourth-century Christian writer St. Jerome, with characteristic attention to the salacious details, claimed Lucretius went insane after drinking a love potion and committed suicide when he was forty-four years old.

imagine how dangerous and powerful they must have seemed to a society that was entirely controlled and ordered by the Christian Church. When it came to religion, *De rerum natura* was implacable: "All organized religions are superstitious delusions . . . [and] are invariably cruel." For Epicureans, "the highest goal of human life is the enhancement of pleasure and the reduction of pain"—a concept that directly contradicts the Christian message that misery in this world earns you joy in the next one.[3] Christian writers twisted this to characterize Epicureans as dissolute and immoral, only interested in their own base desires. In many senses, the poem reads as a manifesto for modern science, and it is so far-sighted we still have not fully understood or explored all the ideas it discusses. Poggio had a copy made in the monastery library and sent it to his friend, Niccolò Niccoli, who beautifully transcribed his own copy, a process that took fourteen long years. Poggio's requests for its return became increasingly fraught and, once he finally received it, copies began to circulate and Lucretius' captivating poetry started to flow through the intellectual networks of Europe, resurfacing all over the continent: in the surreal beauty of Botticelli's *Birth of Venus*, Michel de Montaigne's treatment of sex and death in his *Essays* and Shakespeare's "team of little atomi" who accompany Queen Mab in *Romeo and Juliet*.[4] Fifty manuscripts of *De rerum natura* survive from the fifteenth century, a huge number, demonstrating its popularity—something that was even further enhanced when the work was printed for the first time, in Brescia, in around 1473/4. This was followed by editions in Verona in 1486 and Venice in 1495. The most successful edition of all of these was published in 1500, by the Aldine Press.

De rerum natura was a rare example of a Latin text containing scientific ideas, but scholars who were interested in mathematics, astronomy or medicine knew that they needed to focus on finding Greek manuscripts. For these, they looked to the East, and Venice, with its extensive trading empire, including most of the islands in the Aegean, ports in the Middle East and a lively community in Constantinople itself, was in an unrivalled position to assist. Venetian officials, posted on these islands as magistrates and local governors, experienced first-hand the value of a Greek education, grounded in the study of works of classical literature, philosophy and science in the original Greek. They purchased manuscripts and brought them back to Venice, instituting language lessons as part of the curriculum in the city's humanist schools. They wrote commentaries, had manuscripts copied and built up a canon of Greek texts. In 1463, a chair of Greek was established at Padua University, the only centre of higher education in the Venetian domain and the place where the patricians sent their sons to complete their studies. As the century wore on, the Venetian nobility became increasingly well educated and interested in classical ideas.

Scholars from less impressive backgrounds, native Venetians and visitors alike, joined and enriched the intellectual circles of the elite. Greeks arrived in large numbers during the decades before and the months immediately after the Turks took Constantinople, in 1453. The ancient city was no longer safe for its Christian inhabitants and, when they fled westwards, Venice, with its close connections to the Greek Near East, was the obvious destination. Here, they found fellow Greeks, Greek-speaking Italians and an open attitude towards their homeland. On leaving Constantinople, they took their most valuable and treasured possessions: gold and jewels, obviously, works of art, money, devotional items and books. It was one of the most momentous dispersals of manuscripts in history. Thousands of texts that had been kept safe in the ancient libraries of the city on the Golden Horn were pulled from their shelves and packed into wooden trunks, then trundled on carts down to the port and loaded onto the ships that took their owners into exile in Europe. When they arrived in Italy, humanist scholars were waiting for them, quills in hand, ready to copy, translate and edit them, to produce the best and most accurate versions in order to rediscover the pristine wisdom of the ancient Greeks, uncorrupted by the intervening centuries.

Of all the great scholars who arrived from the East in this period, the most famous was Basilios Bessarion (1403–1472), from Trebizond, on the Black Sea coast. Bessarion had received the best education on offer in Constantinople, then studied Neoplatonic philosophy in the Peloponnese with the renowned sage Gemistus Pletho. A gifted scholar with a talent for diplomacy, Bessarion quickly rose up the hierarchy of the Orthodox Church, and, in the 1430s, he was sent to Italy as part of a delegation to negotiate the reunification of the Eastern and Western Churches. His desire to reunite the Greek and Latin worlds informed every aspect of his long and remarkable career; he was a conduit for the movement of people, ideas and books from Byzantium to Italy, from Greek into Latin. In the process, he saved huge swathes of Greek culture that would otherwise have been destroyed by the invading Ottoman Turks. Bessarion impressed the Italians so much that Pope Eugene IV made him a cardinal, an unheard-of honour for an Orthodox cleric, and this appointment marked his formal entry into Italian life. He moved into an elegant house on the Appian Way, in Rome, the airy rooms and shady loggia of which were thronged with Greek émigrés just arrived from Constantinople, clever young scholars, the great, the good and the learned. Under his generous and inspirational patronage, it became an unofficial academy of humanism, with one of the largest and most valuable collections of books anywhere in Europe. The cardinal set up a scriptorium at his house to supply his own collection and the people who visited him. He worked alongside the scholars, and the margins of many of the manuscripts he owned are peppered with notes written in his own hand. He also produced his own translations and, in keeping with his general desire for unity, wrote a treatise that attempted to harmonize Aristotelian philosophy with that of Plato. His greatest monument, however, is his library, the "most richly-endowed of all the libraries formed during the Renaissance,"[5] home to some of the rarest and most precious manuscripts surviving today.

Bessarion's position as the Pope's envoy meant that he travelled extensively, and, wherever he went, he was on the lookout for like-minded scholars and interesting books. In 1460 he was in Vienna, where he met two gifted astronomers, Georg von Peurbach (or Peuerbach) and Johann Müller (known as Regiomontanus). This meeting was to have far-reaching consequences not only for the men involved, but for the overall development of science. Peurbach was Regiomontanus' teacher, a brilliant scholar who had studied in Italy as a young man and who had refused job offers from the universities of Bologna and Padua. Instead, he returned to his native Austria to teach and to study the stars. His monumental version of the *Alfonsine Tables*, updated with his own observations, was the latest in the long succession of star tables we have encountered. It was Peurbach's magnum opus. Regiomontanus, who matriculated at the tender age of thirteen, was without a doubt Peurbach's brightest and most precocious student, and quickly became his academic partner. They made observations together, noting the appearance of Halley's Comet in June 1456, and endlessly discussed their work, and that of other astronomers. Bessarion commissioned them to produce a new, concise edition of Ptolemy's *Almagest* that could be used for teaching. They started work immediately, but Peurbach died suddenly the following year, aged only thirty-seven, so Regiomontanus continued the work alone, completing it in 1462.

The Epitome was a milestone in the transmission of *The Almagest*. More accessible because of its length (half that of the original), it was extremely clear and well structured, and, as such, "gave astronomers an understanding of Ptolemy that they had not previously been able to achieve."[6] It included theories from a wide range of other astronomers, among them Thabit ibn Qurra, al-Zarqali and the authors of the *Toledan Tables*, and it was based on Gerard of Cremona's Latin translation of Ptolemy's great work, supplemented by details from an original Greek manuscript belonging to Bessarion.[7] This was the very manuscript that Henricus Aristippus had brought to Sicily from Constantinople; today it is in the Marciana Library in Venice. In 1496, just thirty years after Regiomontanus had finished it, *The Epitome* was printed in Venice. It became a standard textbook on the university curriculum and introduced the astronomers of the future—Copernicus, Brahe, Kepler and Galileo—to the brilliance of Ptolemy's system, but also, more importantly, to its flaws. In addressing these specific problems, they came up with innovative solutions that moved astronomy forwards into a new era of understanding.

Bessarion persuaded Regiomontanus to return to Italy with him so that they could continue their work together. Theirs was one of the most important and fertile scholarly relationships of the time; Bessarion taught Regiomontanus Greek, while Regiomontanus shared his extensive understanding of mathematics and astronomy with his patron, writing notes on manuscripts of *The Elements* and *The Almagest* to explain Euclid's axioms and Ptolemy's planetary models. They arrived at the cardinal's house in Rome on 20 November 1461, and Regiomontanus had his first glimpse of his patron's incredible library, which, "with the exception of Pappus . . . contained every major classical source for the renaissance of mathematics."[8] It must have been a momentous event in the young astronomer's life. As a youth, Regiomontanus had been lucky to study under Peurbach, who not only shared his passion for science, but had been to Italy and doubtless brought back books with him that were not available anywhere else in Austria. Arriving in Rome for the first time must have been an extraordinary experience, seeing the classical ruins casually strewn across the city, and feeling the echoes of the ancient past all around. Nor would Bessarion's house have disappointed: scholars filling the air with intense debate in Greek, Italian and Latin; the scriptorium with rows of desks at which scribes scratched on their quires of parchment and paper; and then the books, hundreds of them, filling presses on the walls, some brand new, some antique. It must have taken Regiomontanus' breath away.

Bessarion's library was not the only remarkable collection in Rome at that time. Between 1447 and 1455, the Vatican Library was transformed by the humanist Pope Nicholas V, a passionate admirer of classical learning and former pupil of Chrysoloras. He lured collectors and scholars from all over Italy to come and study in the Curia, and sent agents to Denmark, Germany and Greece in search of new texts, increasing the library's holding from a paltry 340 volumes to 1,160 by the time he died.* Greek texts were systematically translated by a team that included Poggio and his old manuscript-hunting partner Giovanni Aurispa, Lorenzo Valla, and a host of Greek scholars freshly arrived from Constantinople. Under the guidance of Nicholas V, the Vatican Library was set on course to become the most impressive and valuable repository of texts in the world—today it is home to 60,000 manuscripts and 8,000 incunabula. Bessarion was closely involved with the endeavour, advising and encouraging his friend. The pope and the cardinal knew each other well, brought together as much by their positions in the Church as their passion for learning, mathematics in particular.[9]

For the next few years, Regiomontanus travelled around Italy, often in the company of Bessarion, meeting the great humanists of the day: Leonardo Bruni, Leon Battista Alberti and Toscanelli, among others. He gave a series of lectures in Padua on "All the Mathematical Disciplines"—including Arabic astronomy, the work he had done on Archimedes and, doubtless, much more besides. He visited Venice with Bessarion, when the cardinal was sent there as

---

* By 1475, it contained copies of both Euclid's *Elements* and Ptolemy's *Almagest*.

papal legate, and also Viterbo (where he made astronomical observations), Ferrara and possibly Florence, too. He also found the time to write a groundbreaking work on trigonometry. By 1467, however, he had said farewell to the cardinal and returned to Austria, perhaps pulled back by a desire to share the knowledge he had gained with his countrymen, as his mentor, Peurbach, had done.

Bessarion had visited Venice many times and had fallen under the spell of the city on the lagoon. As legate, he had stayed at the Benedictine monastery on the island of San Giorgio Maggiore, from where he could gaze across the pale water of the lagoon towards the piazza and the new facade of the ducal palace, sparkling in the sunshine. But it was not only Venice's beauty that he admired. He was impressed by *La Serenissima*'s unique system of government, and was beguiled by its affinity with his homeland. In 1468, the cardinal was sixty-five, and he was beginning to think about what would happen to his growing collection of books when he died. The Vatican, so close to his house in Rome, was an obvious choice. He also considered Florence, home to so many of the leading lights of the Renaissance, who were pioneering the use of mathematics in the arts. Brunelleschi's octagonal dome, inspired by classical architecture and made possible by applied geometry—and massive hoists—had recently been completed, while his rediscovery of linear perspective was transforming the way artists depicted the world. But Florence was already blessed by wonderful collections of books, the most famous being those founded by the ruling Medici family: a public library in the monastery of San Marco and a private one in their household. Venice had nothing similar, and Bessarion therefore determined to bequeath his precious collection to the city, writing to Doge Cristoforo Moro:

> As all the peoples of almost the entire world gather in your
> city, so especially do the Greeks. Arriving by sea from their

homelands they debark first at Venice, being forced by necessity
to come to your city and live among you, and there they seem
to enter another Byzantium. In view of this, how could I more
appropriately confer this bequest than upon the Venetians to
whom I myself am indebted and committed by obligation because
of their well-known favors to me, and upon their city, which
I chose for my country after the subjugation of Greece and in
which I have been very honorably received and recognized.[10]

Once the terms were agreed, he signed the Act of Donation.
In return for "nine hundred excellent volumes in Greek and Latin,
worth about 15,000 ducats,"[11] the Venetian state would provide a
suitable library building to house the books, make them accessible
to "students of all nations"[12] and ensure they were not taken out of
the city.

In the late fifteenth century, Venice was at the peak of her mercantile supremacy. It was said that anything you dreamed of could
be bought there. According to one visitor, the Piazza San Marco was
"the market-place of the world."[13] It must have been intoxicating for
travellers arriving from Northern Europe, assaulted by the crowds,
the hawkers, guides, quacks and beggars, the stench of the canals
and the scent of spices, the laments of the gondoliers and the constant slapping of water on stone. Every year, a glorious fair filled the
piazza for two riotous weeks. Thousands flocked to gawp at merchants from the four corners of the globe showing off their exotic
wares, and local craftsmen selling shimmering mirrors, ruffled lace
and delicate glassware that gleamed with every colour of the rain-
bow. The city was the world's capital of luxury, purveyor of fashion
and queen of commerce. The Piazza San Marco was always full of
traders, but the real centre of business in the city was the Rialto. In
the eleventh century, the government set up offices governing eco-
nomic affairs and the market was enlarged. A hundred years later, it

had evolved into a sprawling bazaar, with colonnades housing specialist shops around the edge, private banks, warehouses and docks (*riva*) on the canal, where goods were constantly being loaded and unloaded.

The incredible wealth generated by Venice's commercial success funded the building of elaborate palaces along the banks of the Grand Canal, known as *casas* (shortened to *ca'* in the staccato Venetian dialect). These were both the private homes and public offices of the great patrician dynasties, where they ate with their children and where they made deals with merchants from abroad. Appearance was everything, there was no need to build fortified castles and dig moats in a city surrounded by water. This gave Venetian architects and builders the freedom to focus entirely on beauty and form. Patrician families competed to have the most splendid facades with extravagant loggias and elaborate carving. The Bon family took this rivalry to its garish conclusion by having the exterior walls of their palazzo gilded and encrusted with jewels, earning it the nickname the *Ca' d'Oro*—the House of Gold.

Decadence and extravagance had become hallmarks of patrician life, but so had culture and scholarship, both of which they pursued unashamedly, inviting the brightest scholars to come and work in the city. Young Venetian nobles studied at home with private tutors and at the recently founded schools, which had adopted a humanist curriculum, teaching Greek alongside the more traditional subjects, like rhetoric and logic. And when a school of philosophy was founded at the Rialto, in 1397, mathematics was integral to its curriculum. These establishments prepared their pupils to study at the University of Padua, which, by the mid fifteenth century, had an excellent reputation for teaching both the arts and medicine. Scholars travelled from all over Europe to study Aristotelian natural philosophy, the first few books of Euclid's *Elements* and parts of *The Almagest*. In medicine, the major textbooks were by Galen and Hippocrates, in

. A page of a manuscript of the *Pantegni*, thought to have been produced in the eleventh century, in the scriptorium of Montecassino, under the supervision of Constantine the African himself. It is thought to be the oldest medical book in Western Europe.

. (*below left*) Constantine the African lecturing on uroscopy and examining urine samples held up by his students. Urine was a vital diagnostic tool in medieval medicine and Constantine was instrumental in educating Europeans in the art.

. (*below right*) Frederick II pictured with one of his beloved birds of prey in a manuscript of his groundbreaking treatise on falconry.

22. An illustration showing "Saracen (Muslim), Greek and Latin scribes working at the royal court of Palermo.

23. One of the glittering mosaics in the Room of Roger in the Norman Palace in Palermo.

24. Mosaic in the Martorana Church, Palermo, showing Roger II being crowned by Jesus Christ.

25. Roger II's lavishly embellished vermilion silk mantle, made for him by specialist craftsmen in his palace workshop.

26. *(left)* Circular map of the world from a fifteenth-century edition of al-Idrisi's *Book of Roger* (1154). North is at the bottom and south is at the top, as was customary in Arabic maps of the period.

27. *(below)* Sicily, as shown by one of the many maps in al-Idrisi's *Book of Roger*. The "toe" of Southern Italy can just be seen on the left-hand side. Again, north is at the bottom and south is at the top.

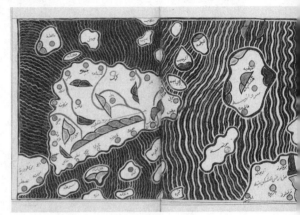

28. Miniature of Adelard of Bath, elegantly dressed in a scarlet robe with an elaborate hat, teaching two students, with a board showing Hindu-Arabic numerals and diagrams on the wall behind him.

29. A Christian (*left*) and a Muslim (*right*) enjoying a game of chess, which, like so much else, originated in northern India and gradually spread through the Arab world and then into Europe during the medieval period.

diogonalis quadranguli cui⁹ latera ſūt diuerſitates aſpectus in
longitudine & latitudine. Diuerſitas aſpectus lunę ad ſolē eſt
exceſſus diuerſitatis aſpect⁹Lunę ſup diuerſitatē aſpectus ſolis
Si uera cōiunctio luminariū fuerit inter gradum ecliptiçę aſcē
dentē & nonageſimū eius ab aſcendente: uiſibilis eorū cōiun/
ctio pręceſſit uerā.Si autē inter eundē nonageſimū & gradū oc
cidēte fuerit uiſibilis uerā ſequeť. Sed ſi in eodē gradu nona
geſimo acciderit tūc ſimul uiſibilis cōiūctio cū uera fiet nullaq;
diuerſitas aſpect⁹in longitudine cōtinget. Nonageſim⁹ nāq;
gradus ecliptiçę ab aſcendēte ſemp ē in circulo p zenitb &ɓpo/
los zodiaci ṗcedēte.   Latitudo lunę uiſa ē arcus circuli magni
THEORICA ECLIPSIS LVNARIS.

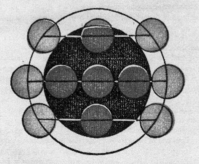

30. Diagram showing the phases of the moon—the first tri-colour printed diagram produced by Erhard Ratdolt in his 1485 edition of Johannes de Sacrobosco, *De sphaera mundi*.

31. Portrait of Luca Pacioli shown demonstrating one of Euclid's theorems to an unidentified young man. The red book on the desk is thought to be a copy of Pacioli's magum opus, the *Summa de arithmetica*.

32. Opening page of one of Ratdolt's presentation copies of Euclid's *Elements*, showing the letter of dedication to the Doge in gold ink.

3. (*top left*) Woodcut of an armillary sphere from a 1543 edition of Regiomontanus' *Epitome of the Almagest.*

4. (*top right*) A miniature portrait of Poggio Bracciolini in a manuscript of his *De varietate fortunae.*

5. The manuscript of *The Almagest* in Greek brought to Sicily by Henricus Aristippus, and then acquired by Cardinal Bessarion.

36. Pages from the Latin translation of *The Almagest* produced in Sicily, which was later owned by Coluccio Salutati.

37. This beautifully illuminated manuscript shows Galen in an apothecary's shop with a scribe and an assistant, who is busy pounding ingredients with the pestle and mortar, while canisters of herbs and drugs line the high shelf behind.

addition to certain sections of Gerard of Cremona's translations of Avicenna's *Canon* and Razi's *Liber continens.*

Padua had come under Venetian rule in 1405, and, from 1407, Venice's young men were forbidden to study anywhere else in Italy, a classic example of the state's need to control. But this also had a positive effect, ensuring a constant exchange of ideas between the two places. Padua's influence encouraged Venetian intellectuals to focus on the sciences, in contrast to the other capitals of Renaissance culture, Florence and Rome, where art, architecture, philosophy and literature reigned supreme. As merchants and navigators, the Venetians were pragmatists, interested in applying scientific ideas to the practical problems they encountered in navigation, accounting, boat building and craftsmanship. As trade increased and transactions grew more complex, merchants needed increasingly sophisticated levels of mathematical knowledge. The most widespread application of mathematical theory to everyday life was Fibonacci's theory of arithmetic, as set out in his *Liber abaci*. It was taught in schools throughout northern Italy, equipping young men with accounting skills, basic algebra and elementary geometry. According to the writer Giordano Cardano, Fibonacci's more challenging ideas—the complex algebra and the number theory of his famous sequence—lay dormant for three centuries, until a young Perugian scholar called Luca Pacioli (1447–1517) stumbled upon a manuscript copy of the *Liber abaci* in the library of San Antonio di Castello, in Venice. Pacioli took up these ideas and included them in the *Summa de arithmetica*, the huge mathematical compendium he assembled, bringing them to the attention of later generations of scholars.

Pacioli was in the city working as a tutor to a noble Venetian family, attending lectures at the Scuola di Rialto in his spare time. Around 1470, he left for Rome, and eventually became a Franciscan friar, dedicating himself to a life on the road. But he seems to have spent as much time teaching maths as he did preaching the Word of

God. Petrarch, Boccaccio and other early humanists had promoted the Italian language, and Pacioli continued this, calling for translations from Latin into Italian and for new works to be written in the vernacular. He believed that everyone should have the chance to learn geometry and arithmetic, and also supported the adoption of Hindu-Arabic numerals. As a result, he holds a unique place in the spread of mathematical ideas during the Renaissance. Pacioli was the ultimate wandering scholar. His itinerant lifestyle made him one of the best-connected men of his time, at home in every noble court and university in northern Italy. He stayed with the architect Leon Battista Alberti in Rome, visited Naples and, in Florence, lived with Leonardo da Vinci for a time. He holds the important, if unexciting, epithet, "the father of accounting," thanks to his lucid explanation of the "method of Venice" (now known as double-entry bookkeeping), which was used by generations of merchants. Pacioli's mathematical interests were especially broad; he wrote on arithmetic, was an expert on Euclid and regularly lectured on *The Elements*, producing new editions in both Latin and Italian. He brought all his knowledge together in his magnum opus, the *Summa de arithmetica*, written in Italian so as to be accessible to as many people as possible, and published in Venice, in 1494. It is a masterful compilation of practical subjects, like weights and measures, with more theoretical areas, like algebra and geometry. It is not an original work, but it had a huge impact in the following century by providing scholars with a useful summary of the theories of Euclid, Khwarizmi and Fibonacci. Pacioli emphasized how important the study and application of mathematics was for professionals such as surveyors, carpenters, engravers and architects, bringing the practical and theoretical elements of mathematics together for the benefit of all.

The dissemination of information on the kind of scale envisioned by Pacioli had only recently been made possible by the arrival of the printing press, which had made books accessible and afford-

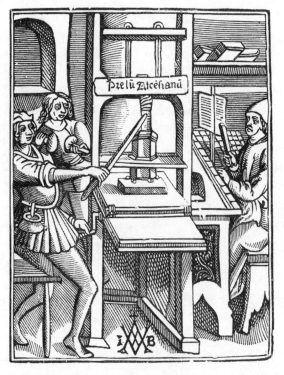

25. A woodcut illustration of an early printing press.

able (relatively speaking) to a much larger section of society. In the mid 1430s, a young gem-cutter in Strasbourg, Johannes Gutenberg (1400–1468), had designed a revolutionary way of producing books, a technology that would alter the course of history. Using his metalworking skills, he cast hundreds of letters in a special alloy of tin, copper and antimony, which he then arranged into words and paragraphs inside a frame. He coated the surface of the letters with ink and fixed the frame, face down, into the wooden press he had built, based on the standard design for a cider press. After sliding a sheet of paper into the press, he pulled the lever down, imprinting the ink on the letters onto the page and creating the first sheet of printed text in the Western world.* Gutenberg spent several years perfecting his design, but, by 1450, he had returned to Mainz, his birthplace, and opened the first printing press. Five years later, he printed about 180 copies of his most famous work, the Gutenberg Bible, in just a few weeks. Scribes would have taken years to complete the same task, and it was this enormous increase in the speed of production that made the printing press so significant. The groundbreaking brilliance of his invention was not lost on either Gutenberg or his contemporaries, and the news travelled quickly. The future Pope Pius II admired samples from a printed Bible in Frankfurt, and wrote to friends in Italy with an astonished description of how clear the type was. Men were trained to build and operate the press, a skill that spread quickly, at first within Germany and then in other countries, especially Italy.[14]

The Venetians, always on the lookout for an opportunity, immediately saw the potential of the printing press and, in 1469, issued Johannes of Speyer with a privilege (monopoly) to print books in Venice, declaring that, "this peculiar invention of our time, although unknown to former ages, is in every way to be fostered and advanced."[15] Johannes died the following year, and his monopoly died with him, allowing others, including his brother, Wendelin, to set up presses in the city. Within three years, more than 130 editions had been published; over half were of classical literature or grammars, the next largest group was religious texts, the remainder were books on law, philosophy or science. There was no doubt, they had backed a winner. In truth, the city had all the conditions necessary for printing to flourish: a large, educated reading public, a well-organized banking sector to provide the finance, an enterpris-

* The Chinese had invented their own version of the printing press in the early thirteenth century.

ing government, an established trade network and, crucially, a reliable supply of paper from the Venetian provinces on the mainland. The paper-making industry was by now well established in Europe, having taken several centuries to make the journey from Baghdad, via Spain, to Italy. Most important of all, the city not only welcomed foreigners, it actively invited them; the first generation of printers all came from Germany, joining the thriving community of merchants who gathered at the Fondaco dei Tedeschi. As a consequence, Venice was soon pre-eminent in printing. By 1500, there were around thirty presses operating in the city, and between 35 and 41 per cent of the total number of books printed before 1500 came from Venetian presses.

Printing took off quickly in Venice, but it was certainly not a job for the faint-hearted. A printer needed a daunting array of skills and expertise: woodwork, chemistry, languages and metallurgy. He had to be an artist, a businessman and a scholar all at once. Then there was the print shop, a noisy, dangerous place, where vats of boiling oil bubbled and handling corrosive chemicals and burning pitch (to make the ink black) were daily occupations, never mind operating the heavy wooden frames of the press itself. Many early printers were unable to deal with the huge technical challenges of the process and the financial outlay necessary to produce the books. Competition was intense, many went out of business and lost everything— only 25 per cent managed to stay afloat for more than five years. Venice's print shops were concentrated in the Merceria, the bustling, narrow streets linking the Piazza San Marco with the Rialto; signs displaying their individual printers' devices swung outside each one, and books were laid out on tables for customers to browse, while the presses clanged in the workshops at the back.

These presses churned out many different kinds of books: classical literature, astrological guides, school textbooks, Bibles, practical manuals and a surprising number of scientific works. When it came

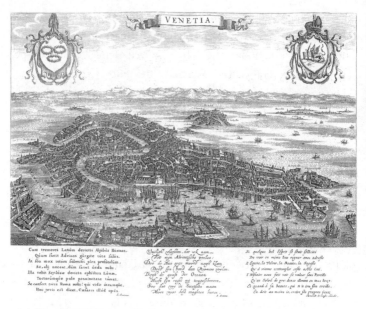

26. A fifteenth-century map of Venice.

to scientific publishing in late-fifteenth-century Venice, the most significant figure was a man by the unprepossessing name of Erhard
Ratdolt. Ratdolt came to Venice from his native Germany in 1475,
just a few years after the first printing house had opened for business. He joined forces with two other Germans and set up a press.
The very first book they printed was Regiomontanus' *Calendarium*,[*]
leading historians to speculate that Ratdolt had worked for the
astronomer in Nuremberg and obtained the manuscript directly
from him. Regiomontanus had returned to Germany in 1467 and,
four years later, decided to print his entire catalogue of mathemati-

---

[*] This was a diary containing astronomical data, the dates of festivals and fasts,
showing when the sun entered the different signs of the zodiac. Ratdolt printed it
in Italian and Latin.

cal texts himself, with a view to establishing scientific standards and consistency. Having secured the backing of a wealthy merchant in Nuremberg, he set up an independent research institution with its own library, printing press, observatory and workshop for producing instruments. This marked a turning point in the role of the scientist in Western Europe. No longer a wandering scholar, reliant on the patronage of nobles and churchmen, Regiomontanus was now completely independent, the first in a line of astronomer-printers who would dominate the subject for years to come. The first book he published was Georg Peurbach's *Theoricae novae planetarum*, a homage to his mentor and teacher.

In 1474, Regiomontanus extended his publishing programme to forty-seven works, including, among others, *The Almagest*, *The Elements* and other works by Ptolemy and Euclid, everything available by Archimedes, Apollonius' *Conics* and other texts from the *Little Astronomy/Middle Collection*, together with some of his own works, such as *The Epitome*. In other words, the full canon of mathematics and astronomy. He also began to publish annual ephemerides (books of astronomical tables, *zij*) listing the positions of the stars and planets, combined with other celestial information, for each day of the year. They have been produced continually ever since, used for navigation, astrology and the study of astronomy. These days, they are generated by NASA, using specially designed software, and are mainly used for navigating spacecraft.

Such an ambitious enterprise would have required a considerable workforce, and while we cannot be absolutely certain, it seems that Ratdolt was one of the young men Regiomontanus employed to help in the press. At this point, not many people had mastered this new technology, and we know from his later publishing career that Ratdolt was interested in astronomy and mathematics, making him the perfect candidate for Regiomontanus' project. Assuming they worked together, it is also likely that Regiomontanus would have

27. First page of Ratdolt's 1482 printed edition of *The Elements* showing the geometrical figures and their names.

told Ratdolt about the wonders of Italy and recommended Venice as a good place to set up a printing press. Most important of all, it would explain why the first book Ratdolt printed in Venice was the *Calendarium*, presumably based on a manuscript he brought with him from Nuremberg.

The *Calendarium* was the first of many astronomical and mathematical books Ratdolt produced. In 1482, he issued the first printed edition of *The Elements*, based on the Adelard/Campanus version. This milestone in the history of Euclid's great work, marking the end of its long journey, from fragile scroll in ancient Alexandria to printed book in Renaissance Venice, is also a key moment in the history of mathematics, and the history of printing: thanks to Ratdolt's ingenuity, it was the first time that diagrams were printed in a text. He produced two special presentation copies, printed on vellum, with a dedicatory letter to the doge in gold ink. In the letter, he explained that he had not understood why such a seminal book had not been printed earlier, until he realized how challenging the task of producing the diagrams was. He solved this problem by making 420 separate woodcuts, which he printed in the specially designed, extra-wide margins of the book, retaining the decorative border on the title page and large initials at the beginning of each chapter that embellished the manuscript version. Printed books had yet to develop their own style and were still designed to look as much like manuscripts as possible. Ratdolt's edition of *The Elements* stands alongside the other seminal copies of the book: the papyrus fragments found at Oxyrhynchus and the outstanding copy purchased by Bishop Arethas in 888. It is a monument to Venetian printing, to the transmission of mathematical knowledge and to Erhard Ratdolt.

The Elements was reissued and printed several times in the next few decades. In 1505, a new Latin translation based on a Greek manuscript was printed in Venice and, three years later, Pacioli returned to the city to prepare another new Latin translation of the text for press, which was based on the Adelard/Campanus tradition, but with amendments and corrections. In 1533, the first edition in Greek was published, and a decade later it appeared in Italian—other European vernaculars followed a few years later. Ratdolt became one of Venice's most successful and respected printers. In 1485, he published eleven titles and built on his pioneering achievement in printing figures and diagrams, inventing a method that enabled him to use three different coloured inks together on the same page. He showcased this new technique in a collection of astronomical texts, beautifully illustrating a description of lunar eclipses with a diagram showing every phase of an eclipse's progress. Ratdolt was also responsible for the first "modern" title page, the use of Arabic numerals to date his books and the issuing of type-specimen sheets and lists of errata. News of Ratdolt's successes must have reached his home town of Augsburg, because the bishop wrote asking him to return and put his expertise in the service of his own people. So Ratdolt packed up his types and his woodcuts, his wife and his children, and went back to Augsburg, where he spent the rest of his career mainly printing religious books. He has largely been forgotten, given little credit for his many remarkable innovations, and completely dwarfed by the other great innovator of the first age of printing in Venice: Aldus Manutius.[*]

The major difference between Aldus and his fellow printers was that he was a serious scholar in his own right, whereas they were generally craftsmen, albeit with intellectual interests. This was important because every book that was printed was produced using at least one manuscript (known as an exemplar), if not several; producing a definitive edition of a text required specialist skills and knowledge. Once the text had been prepared, the type was set by compositors, who would sit on high stools with the manuscripts in front of them. It was a time-consuming process and a complex one. Manuscripts were often difficult to read, and there was no standardization of spelling or fonts, so compositors had to be both intelligent

---

[*] There is a plethora of studies on every aspect of Aldus Manutius and the Aldine Press, but only two that focus on Ratdolt.

and well educated—if they made a mistake, it would affect the value of the resulting book. Identifying the original manuscripts used by the printers to create the new printed editions is extremely difficult, and very few have been found. Many must have been thrown away, no longer needed now that several hundred printed copies were available, and they were most likely damaged and ink-spattered after weeks in the bustling, dirty environment of the printing house.

There was certainly no shortage of collections in Venice to supply the presses with manuscripts; collectors, scholars and printers often worked together to produce printed texts. With his linguistic skills and his wide-ranging knowledge, Aldus Manutius was able to bring all these strands together. He was the colossus of Venetian printing, creator of italic script, small-format books, clearly legible Greek fonts, the semicolon and a host of other innovations, and many people credit him with inventing books as we know them today. Manutius studied at the University of Rome before moving to Ferrara to learn Greek and work as a tutor to young noblemen. He seemed to be following the traditional path of an Italian Renaissance scholar, but, in 1489, he suddenly changed course and moved to Venice. Five years later, he founded the Aldine Press.

Aldus settled quickly into life in Venice, "with the effortless grace of one who knows something of his worth."[16] With his formidable intellect, his easy charm and his boundless enthusiasm for scholarship, he was soon a key member of the circle who orbited around Giorgio Valla, the greatest mathematician in Venice at that time, and the owner of one of the most important collections of manuscripts. Valla's major contribution to the transmission of scientific ideas lay in the enormous compendium of sources of classical mathematics and philosophy that he compiled from his own manuscripts, called *De expetendis et fugiendis*. Manutius published it in 1502, but by then Valla had been dead for two years, and his library was no longer in Venice.* It was an extremely influential book, providing scholars of the next generation with a wide range of well-structured, clearly translated, accessible scientific material, used regularly as a reference guide and, in many cases, the only printed edition of a source text. Valla had been a public teacher of mathematics, but he had also presided over a circle of Greek scribes and lectured on architecture and poetry. Two of his pupils went to Messina, in Sicily, to perfect their Greek. They returned in 1494 with a Greek language guide, which Aldus used in combination with Chrysoloras' grammar to produce one of the first books he published—a definite statement of his intent to promote the study of Greek.

Another close friend of Valla's, indeed the man who had brought him to Venice, was Ermolao Barbaro. Ermolao was bilingual in Greek and Latin, and this enabled him to build upon the books he inherited from his father and grandfather, which were housed in the sumptuous family palazzo, just along the Grand Canal from the Ca' d'Oro. He has left us an idyllic insight into the summer routine of a wealthy intellectual: "the morning was spent in intensive study of Aristotle and the Greek orators or poets: then came a light lunch of broth, eggs and fruit; afterwards, more relaxed reading or dictation, followed by conversation with any friends who cared to call for a literary or philosophical discussion; finally, a supper of roast game, a stroll in his botanical garden to ponder the herbal lore of Dioscorides, and so to bed."[17] These ponderings resulted in Barbaro making his own Latin translation of the *De materia medica*, but he is most famous for his scathing attack on the inaccuracies of Pliny's *Natural History*.

Unlike many printers in Venice at the end of the fifteenth century, Aldus Manutius was neither German nor French, but Italian. He founded the Aldine Press in 1494–5, at a time when chaos was

---

* The majority of his collection is preserved today in the Estense Library, in Modena.

sweeping across mainland Italy in the form of the invading French army and a virulent outbreak of plague. Venice didn't escape the plague, which killed thousands, but it did avoid the French, who were not equipped to sail across the lagoon and attack the city. This extraordinary stroke of luck allowed Venice to overtake Florence as the intellectual capital of Italy. Florence had slid into violence and then repression under the influence of the fanatical monk Savonarola, who stifled intellectual enquiry in the city, and many scholars fled, some of them to Venice. As the city's most prominent printer, Aldus played a leading role in Venice's rise. His success was down to several factors. First, he was a gifted individual, both as a scholar and as an entrepreneur. Second, he was lucky; he arrived at exactly the right moment, and identified a gap in the market—printing in Greek. Having designed (with the help of his collaborators) a series of elegant Greek fonts (one of which was based on Chrysoloras' handwriting), he began to print classical texts in their original language, and, in doing so, fulfilled the ultimate humanist ideal of making the pristine knowledge of the ancients accessible to a contemporary audience, uncorrupted by translation. His print shop, first in San Agostino and later in the Merceria, became the intellectual heart of the city. Every day, an endless stream of scholars arrived to debate the latest issues—in Greek (there were fines for speaking any other language)—and to prepare texts for the press. The leading figures of the European "Republic of Letters" all came to pay their respects: Erasmus arrived in January 1508, to oversee the publication of his *Adages*; the German humanist Johann Reuchlin visited a few years earlier, while Thomas Linacre came all the way from England. This did not make the Aldine Press an easy place to work. In 1514, the year before he died, Aldus wrote, "Apart from six hundred others, there are two things in particular which continually interrupt my work. First, the frequent letters of learned men which come to me from every part of the world . . . then there are the visitors who come . . . and sit around with their mouths open."[18] Being at the centre of the intellectual world had its disadvantages.

For a long time, historians assumed that the printed editions which rolled off Manutius' press were created using Bessarion's collection of manuscripts, which he had donated to Venice twenty years previously, but this does not, in fact, seem to have been the case. The books had arrived in two loads. The first, from Rome, in 1469, was carried over the Apennines in thirty cases by a convoy of mules. The rest were sent from Urbino, where Bessarion had left them under the care of Duke Federigo da Montefeltro, a passionate mathematician and patron of education. On arrival in Venice, they were stored in a room in the Doge's Palace, still in their packing cases. They remained there, rotting gently, until 1531, when they were finally taken out of the boxes and put on shelves in a room above the doors of the basilica. It would be another thirty years before the library building, promised to Bessarion in exchange for the bequest of his books, was finally constructed and the Biblioteca Marciana was born. So, in one of the sadder ironies of the history of printing, when Manutius was publishing his seminal Greek editions of Aristotle, Aristophanes and the rest, there were exemplary copies of their works, in the original Greek, lying in boxes on the other side of the city, but beyond his reach.

Manutius' greatest talent was his ability to market his books: he was "one of the first to take full measure of how the world of books had changed in the last twenty-five years of the fifteenth century, and to devise a strategy of marketing and publicity that took account of these changes,"[19] and he led the field. From 1502 onwards, the Aldine "dolphin and anchor" device was central to this strategy; stamped onto the title page of all his editions, it was the guarantee of Aldine quality, imbued with an aura of authority and excellence— possibly the first example of successful branding. The extent to which it was forged by other printers is clear evidence of its power.

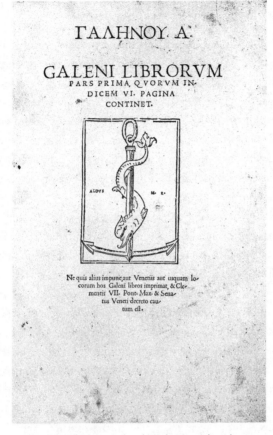

ΓΑΛΗΝΟΥ Α΄

GALENI LIBRORVM
PARS PRIMA, QVORVM IN-
DICEM VI. PAGINA
CONTINET.

ALDVS   M· R·

Ne quis alius impune,aut Venetiis aut usquam lo-
corum hos Galeni libros imprimat, & Cle-
mentis VII· Pont· Max· & Sena-
tus Veneti decreto cau-
tum est·

28. The Aldine "dolphin and anchor" device on the title page
of the 1525 edition of Galen's *Opera omnia*.

The Aldine Press's most remarkable publication was the complete works of Aristotle, in Greek—a massive five-volume undertaking that involved scholars from all over Europe sourcing the
necessary manuscripts and editing the final version. Thomas Linacre, the English humanist scholar, was instrumental. During his stay
in Venice in the 1490s, he helped Aldus edit the edition, and the

copy he took home with him, printed on vellum, is in New College Library, in Oxford, today. Each volume is neatly inscribed with his name, "Thomae Linacri." For the first time since antiquity, the full breadth of Aristotelian philosophy was accessible—to those who could afford it, and who knew Greek. But such ambitious publications were expensive to produce and didn't make much money. Aldus soon realized that he would have to diversify his programme to include books that were more attractive to a wider audience, and that, since very few people could read Greek, they would have to be printed in Latin or Italian. Humanist ideals were all very well, but he had to keep the press afloat.

The last years of the fifteenth century and the first half of the sixteenth saw the publication of many philosophical and literary texts in Greek. The same cannot be said of medicine. The manuscripts simply weren't available to printers and very few doctors could read Greek, so it is doubtful that printed editions of medical texts would have been very successful. Needless to say, Bessarion's collection contained a good selection of treatises by Galen, but, as we have seen, they were languishing, unused, in the Doge's Palace. Medical teaching was well established on the university curriculum, based on the *Articella* texts. Many people considered these to be sufficient, but, behind the scenes—or, rather, within the walls of private collections—things were changing. In the course of their searches for manuscripts, humanist scholars had inevitably discovered previously unknown works by Galen (there is an apparently almost inexhaustible supply—new ones are occasionally unearthed even today, almost 2,000 years after they were written).[*] Examining these new treatises made scholars aware of aspects of Galenic medicine not present in the Arabic and medieval traditions, and these stimulated

[*] In early 2005, a French research student discovered a treatise by Galen called *On the Avoidance of Grief* at a monastery in Thessaloniki.

new avenues of research and highlighted inconsistencies in those
traditions. When they did encounter a theory that was manifestly
incorrect, their reverence for Galen was such that they blamed the
scribes who had copied the texts. At this point, any notion that he
might have been wrong was inconceivable. Ironically, the new texts,
and translations of existing ones, eventually forced scholars not only
to accept that Galen had made many fundamental mistakes, but to
replace his theories with new ones of their own.

However, in the early years of the sixteenth century, Greek

29. A page of Galen's *De methodus methendi* (*On the therapeutic method*) from a
fourteenth-century manuscript in Greek that was owned by the Barbaro family, who
purchased it from a Cypriot doctor in the second half of the fifteenth century. In
1517 the English scholar Thomas Linacre translated the text into Latin and published
it in Paris.

thought, specifically the work of Galen and Dioscorides, was still idolized and energetically promoted—especially by a physician-collector whose pedigree included teaching and studying at Padua, Bologna and Ferrara. During his long career, Niccolò Leoniceno (1428–1524) had assembled an extraordinary library of Greek medical and scientific manuscripts. According to Professor Vivian Nutton, it was "more extensive than any other known before or since, and distinguished not only by its sheer size, but also by the rarity of its contents."[20] Using his collection as a springboard, he launched an attack on the centuries of mistakes, misinterpretations and scribal errors made in the transmission of medical science, especially in works on disease and medicines. He included Roman writers in his tirade, particularly Pliny the Elder, who, he claimed, had corrupted Dioscorides' *De materia medica* by misidentifying plants and filling it with inaccuracies. The upshot of this was renewed veneration of the original Greek source, and, in 1499, Manutius, a close acquaintance of Leoniceno, published the first Greek edition of *De materia medica*. Dioscorides' work therefore reached a much wider audience than any of Galen's works had at the time. Apothecaries—indeed, anyone with an interest in botany and botanical drawings—would have been keen to own a copy. The Aldine edition was a great commercial success, despite the complexity of producing woodcuts for the illustrations of the plants.

Leoniceno lent manuscripts not only to Manutius, but to other printers, too. However, it was not until his death, at the grand old age of ninety-six, that the full wealth of his Greek Galenic collection became available. In 1525, the Aldine Press printed the *Opera omnia* in Greek, a huge and costly undertaking that was only possible because Venice was not at war, so that, "a large supply of metal that might otherwise have been used in the Arsenale to make cannon"[21] could be purchased and used to cast the enormous quantity of type necessary to produce the work. This was naturally reflected in the high retail price—thirty florins, or gulden, in Germany (one third of a Nuremburg doctor's annual stipend), fourteen scudi in Rome—it was only affordable to the wealthy. As usual, Galen's extreme verbosity worked against him. Even attempts to publish small parts of his output could be risky; in 1500, a Venetian printing house was bankrupted by their Latin edition of the two Galenic treatises from the *Articella*. The Aldine *Opera omnia* was not a particularly distinguished edition, but it did make many new aspects of Galen's work available to the medical community, especially after translations in Latin, based on the printed Greek versions, began to appear. The interplay between philosophy and medicine, Galen's pharmacology and his debt to Hippocrates all became much clearer, as did his ethical standards and his ideas about correct medical practice. As a result of this, medicine began to change and develop in interesting ways. In Padua, teachers now combined theoretical and practical study, making a closer link between the lecture hall and the sick room, while, encouraged by the publication of Galen's work on veins, arteries and nerves, dissection and anatomy became increasingly important, paving the way for the discoveries of the great Flemish anatomist Vesalius in the 1540s.

Books produced in Venice were sold in the Merceria to an endless stream of locals and visitors, but they were also packed up, loaded onto boats and sailed up the Po, on the first stage of their journeys to other Italian cities, and on to Germany, France, Spain and England. This vast distribution network carried books across the continent, into bookshops and homes, making them more accessible than ever before. With the invention of printing came standardization, sources of information became much more uniform and, provided the printer had done his job properly, accurate. The price of books fell significantly as production increased and the marketplace for them grew. Smaller octavo-format books were instrumental in the price decrease; using far less paper, they were cheaper to produce—paper was expensive, and represented 50 per cent of printers' costs. At first, octavos were exclusively devotional books, until Aldus began printing classics in this size, and, while his books were always quite expensive, other printers took up this innovation and sold them for less. By the end of the sixteenth century, even literate artisans could afford to buy books, and, by then, there were also many more books on offer in vernacular languages.

By 1500, the "compact, bounded and ordered" universe of the previous century had begun to split apart.[22] The world was expanding, pushing the frontiers of knowledge and forcing humankind to reimagine their place within it. It was no longer possible to believe that classical thinkers held the key to everything, that ancient books could provide all the answers. The new, printed editions of Euclid, Galen and Ptolemy were important in disseminating their ideas, but they were also instrumental in highlighting their flaws. In the sixteenth century, scientists would focus on correcting these inaccuracies and replacing them with new theories based on detailed investigation into the natural world, paving the way for the extraordinary discoveries of the seventeenth-century scientific revolution.

At the close of the fifteenth century, the great works that we have followed from antiquity had all appeared in printed editions; their legacy was secure. What followed was a period of assimilation and correction, rediscovery and rescue. That vital work of reassessment would enable the scientists of the next generation to build on the ideas of Euclid, Galen and Ptolemy, and all those who had preserved their writings for a millennium, and revolutionize astronomy, mathematics and medicine.

NINE

1500 and Beyond

LOOKING DOWN ON the map of knowledge in 1500, the picture has changed dramatically. Cities have risen and fallen, new societies have developed across the Mediterranean world. In 500, centres of learning were shutting down, intellectual life diminishing. A thousand years later, the opposite is true. In Europe, education is widely available again, not to everyone, but there are schools, tutors, universities: a budding tradition of learning on offer to wealthy, interested young men—and a few women, too. They have the chance to become members of the growing "Republic of Letters" and to contribute to the development of knowledge.

Europe has emerged from a century of profound change. New worlds have been discovered, bursting with exotic plants and animals. Galleys laden with gold and silver are making their way back across the Atlantic, bringing untold wealth to Europe. Old boundaries have been swept away and the maps have been redrawn. The printing press has transformed communication. By 1500, there are presses in 280 towns in Europe, which have produced in the region of "20 million individual books."[1] Knowledge is cheaper, more accessible and more widely available than ever before. In the coming decades, the printing press will help facilitate a religious revolution and acceleration in scientific progress.*

While Christian Europe flourished, the Muslim Empire frac-

---

* It is, however, worth noting that many scientific works were still circulating in manuscript form during this period.

tured and contracted. By the mid sixteenth century, it had split into three separate political entities. In the ensuing tumult, there was neither the time nor the money to fund ambitious programmes of mathematical exploration, astronomical observation or medical research. The two great discoveries of the fifteenth century, the New World and the printing press, were disastrous for Islamic fortunes. European voyages of discovery opened up new trade routes by sea that bypassed the Middle East, depriving it of commercial opportunity. The ancient Silk Roads, which had conveyed such great riches over the centuries, grew quiet and desolate. As printing presses opened in towns across Germany, France, Italy and England, in the Muslim world people remained suspicious of this new technology and struggled to design moveable type for Arabic, with its whirling diacritics and myriad variations. For this, and many other reasons, it took them centuries to adopt the printing press, putting them at a huge disadvantage in the dissemination of knowledge. The focus of the scientific enterprise moved, swung northwards to Italy, to France and Germany, and to England. The Islamic world began to consume, rather than produce, scientific information.

Given these circumstances, coupled with increasing religious conservatism, it is perhaps unsurprising that the pursuit of knowledge in the Muslim world began to wane. But it is less easy to understand why the legacy of Islamic science has been largely forgotten in Europe. Given the remarkable contribution they made, scholars like al-Khwarizmi and al-Razi should be household names, like Leonardo da Vinci and Newton, but, even today, few people in the Western world have heard of them. How did this happen? Part of the blame must lie with the humanists, whose idolization of Greek science led them to disregard many scientists of the intervening period. Medieval translators were also guilty of "Latinizing" the books they translated and failing to credit the original Muslim authors. And, as Europe grew in wealth and power, and began to build empires, it gained the cultural upper hand, too. As a result, a narrative developed that marginalized Arabic learning and pushed it back into the past.

This process is embodied by a dramatic act of iconoclasm that took place in 1527. The radical German scholar Paracelsus publicly burned his copy of Avicenna's *Canon* as part of his call for medical students to turn away from the "little books of men" and turn instead to "the great book of nature."[2] Paracelsus occupied an extremist corner in a wider movement promoting new approaches to learning, which involved practical observation of the natural world that "would free humanity from subjugation to the dead hand of past authority."[3] But, of course, a good scholar needs both, and the point Paracelsus missed was that one can only reconstruct an intellectual theory from within. This was something that Andreas Vesalius came to appreciate during his years studying Galenic anatomy. It took him a long time to accept that the legendary physician could be wrong. His epiphany finally came when he noticed that Galen described an extra vertebra, one that was present in apes but not in humans. From this, he realized that Galen had never dissected human bodies, only those of pigs and apes; his own anatomical knowledge, based on extensive examination of cadavers, was therefore superior. Rigorous observation of the natural world had triumphed over ancient wisdom.

Vesalius published his findings in 1543, in a book titled *De humani corporis fabrica*. It was a transformative moment in the study of anatomy. This lavish book interspersed detailed diagrams and illustrations throughout the text, based on woodcuts that he had had specially made in Venice and carefully transported over the Alps to Basle, where the book was printed. This was a milestone in scientific printing, the embodiment of Vesalius' desire for clarity and precision in communication, something he had developed during the years he spent editing and preparing Galenic texts for the press. To be use-

30. A woodcut illustration of Galen performing the dissection of a pig on the title page of the 1565 edition of his *Opera omnia*.

ful, scientific knowledge had to be accurate, and nowhere was this more pertinent than in medicine. "One wrong word may now kill thousands of men," Rabelais noted solemnly as he edited Hippocratic texts for print in 1532.[4]

The very same year that Vesalius published *De humani*, Georg Joachim Rheticus, a young professor of astronomy, was in the busy German town of Nuremberg, preparing another seminal scientific work for the press. Written by his reclusive Polish mentor, Nicholas Copernicus, *De revolutionibus orbium coelestium* would have a similarly profound influence, albeit in a different timescale. Vesalius' *De humani* was an immediate, runaway success, selling in huge numbers and making him a celebrity in the world of medicine. It had a wide appeal, for both medical practitioners and for artists, and its author was a young and enthusiastic self-publicist. The situation with Copernicus was very different. *De revolutionibus* was never going to be a bestseller. "Typographically dull and formidably technical," its complex, recondite contents were only of interest to a small number of academic astronomers, and its central tenet—that the sun was at the centre of the universe—was controversial, to say the least.[5] Copernicus was reluctant to publish the book, understandably nervous about its reception. He was a very private individual, who, having completed his education at the University of Padua, spent most of the rest of his life working in isolation in Poland. By the time *De revolutionibus* was published, he was an old man, and he died the same year.

Copernicus spent decades studying Ptolemaic astronomy. The major issue that preoccupied him was the slippage that occurred between what the model predicted and the actual movements of the heavens. This became increasingly obvious as time went by and had bothered astronomers for centuries, but, despite numerous attempts,

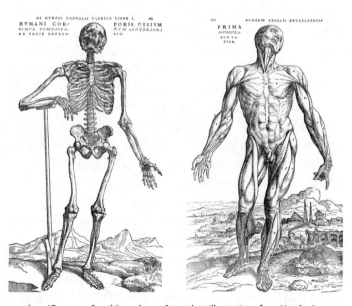

31. and 32. "Bone man" and "muscle man" woodcut illustrations from Vesalius' *De humani corporis fabrica*. As both anatomical diagrams and works of art, they educated and inspired generations of artists and physicians across Europe.

no one had been able to solve it. The discrepancy between the two was especially marked when it came to the spring (vernal) equinox, and accurately predicting this was important because the date of Easter fell on the Sunday after the first full moon after the equinox. Copernicus, as a canon of the Church, was particularly concerned with this problem. His approach was radical. He took Ptolemy's universe and completely redesigned it, putting the sun at the centre, with the planets, now including the earth, orbiting around it. Copernicus retained Ptolemy's geometrical scheme in his new system of the cosmos, providing vital continuity that enabled his fellow astronomers and those of the following generations to develop his ideas effectively. He signalled this aspect of the work by quoting on the frontispiece the phrase that had hung on the doors to Plato's Academy: "Let no one ignorant of geometry enter here." This heliocentric idea had originally been suggested by the Greek astronomer Aristarchus eighteen centuries earlier, and the passage of time had not made it any easier to accept that the earth was simply one of several planets, rather than the privileged orb around which the universe rotated. Worse still, it meant accepting the idea that the apparently stationary earth was, in fact, hurtling through space as it orbited the sun. To say that this shook people up is an understatement.

This new heliocentric cosmos fundamentally altered humanity's place in the universe; accepting it demanded huge psychological and emotional effort, and this was not something that could happen overnight. It totally contradicted religious teaching. "People gave ear to an upstart astrologer who strove to show that the earth revolves, not the heavens or the firmament, the sun and the moon . . . This fool wishes to reverse the entire science of astronomy; but sacred Scripture tells us [Joshua 10:13] that Joshua commanded the sun to stand still, and not the earth," spluttered Martin Luther, on hearing a rumour about Copernicus' theory.[6] Luther was himself a radical, so it's not surprising that the Catholic Church, bastion of tradition and

33. The Copernican universe with the sun at the centre, and the planets circulating it on concentric rings. The earth, "terra," is on the third ring, its tiny moon hovering above it.

conformity, was more horrified still. Truly revolutionary scientific discoveries, ones that initiate paradigm change, are almost always immediately rejected (especially by religious authorities), before being gradually tested, refined and accepted over a long period of time. This was clearly the case with the Hindu-Arabic numerals, which took centuries to be commonly accepted, both in the Muslim Empire and then in Christian Europe, where people were especially concerned about the zero having dangerous, demonic properties.

It was certainly also true of heliocentric theory, the consequences of which were not fully realized until the following century. While *De revolutionibus* was far from being a bestseller, copies were diffused through the scholarly networks of Europe during the decades after its appearance, into the hands of professors and students of astronomy. Unlike Vesalius' work on anatomy, *De revolutionibus* did not contain much new observational data, but, in the following century, Tycho Brahe, a Danish nobleman, built himself an observatory on the island of Hven, in the Kattegat, between the coasts of Denmark and Sweden. He filled it with sophisticated instruments, capable of making far more accurate astronomical observations than had previously been possible. Applying these to Copernicus' new cosmos, he did away with Ptolemy's system of spheres, opening up the possibility of a far more complex and exact representation of the universe. The next step was taken by Johannes Kepler, who, using Brahe's data, reluctantly rejected Ptolemy's theory that the planets moved with regular motion in circular orbits, in favour of elliptical movement—another huge step forward in our understanding of the solar system.

Each of the cities we have visited in this book had its own particular topography and character, but they all shared the conditions that allowed scholarship to flourish: political stability, a regular supply of funding and of texts, a pool of talented, interested individuals and, most striking of all, an atmosphere of tolerance and inclusivity towards different nationalities and religions. This collaboration is one of the most important factors in the development of science. Without it, there would be no translation, no movement of knowledge across cultural boundaries and no opportunity to fuse ideas from one tradition with those from another. The scholars who made this collaboration possible are the stars of the story—the men who set off into the unknown, who devoted their lives to locating, comprehending, preserving and communicating all these extraordinary ideas and theories. Their capacity for wonder, their determination to bring order and clarity to the magnificent chaos of creation, is what drove scientific discovery and kept it alive in the millennium between 500 and 1500.

Throughout this journey, we have tried to glimpse inside the elusive "Houses of Wisdom" of the distant past, where such intellectual activity took place. It has not been easy. There is nothing left of the House of Wisdom in Baghdad, nor the medical school in Salerno, and only dusty ruins or shadowy cathedral precincts remain in the other cities. Even in Venice, which is closest in time to today, the only way of recognizing Aldus Manutius' printing house is a plaque on the wall. Archaeological evidence aside, we know that the books we have followed must have been shelved and read in many different places: royal libraries, cathedral chapels, classrooms, gardens and observatories. By 1500, these places had become more numerous, more various and more visible than ever before, and, during the following century, anatomy theatres and observatories were built, botanical gardens planted, lecture halls and libraries established across the intellectual landscape. In these new sites of academic activity, scholars could work together and make the most of increasingly sophisticated equipment for examining the natural world. Universities played a vital role in education, but they were not usually the location of cutting-edge scientific enquiry.* Their libraries, like royal libraries, were often inaccessible, and their approach to learning conservative. Because of this, scholars set up their own centres of research, which is why many of the great advances of the scientific revolution in the sixteenth and seventeenth centuries took place in private, rather than official, spaces. As the scientific revolu-

---

* Padua University was an exception. In 1595, it became home to the first permanent anatomy theatre, built to replace the wooden construction that Vesalius had used during his time there. Thomas Bodley's library at Oxford University was also open to scholars and, within a few decades of its opening in 1605, had welcomed readers from across Europe.

tion gathered momentum in the late seventeenth century, institutions and societies were founded to facilitate collaboration, and they became the official residences of the developing scientific disciplines.

A collection of books was almost always at the heart of such places, however, and some were so renowned that they functioned as unofficial research institutions. John Dee's house on the River Thames in Mortlake, home to the most impressive collection of scientific texts in late-sixteenth-century England, was visited by a galaxy of scholars and members of the Elizabethan elite, including Elizabeth I herself. They were not only drawn by the books, but also by the huge array of maps, instruments and curiosities Dee owned, and, of course, by Dee himself. They came to plan voyages of discovery, to discuss philosophy, to study historical records, to discover the secrets of alchemy and to try to commune with angels.

During the sixteenth and seventeenth centuries, many aspects of Galenic medicine and Ptolemaic astronomy were discredited and replaced, but this never happened to Euclid. *The Elements* retained its position as the fundamental mathematical text, translated and printed in all the major European vernacular languages and sold in bookshops across the continent. In 1570, the first full English translation was published, a stunning edition with pop-up diagrams, edited—as we noted in the introduction—by John Dee. In the preface, Dee listed all the disciplines to which mathematics could be usefully applied, and emphasized the importance of making it available to as many people as possible. This was a salient feature of the first age of printing. Thousands of books were translated into vernacular languages to cater for the burgeoning numbers of interested readers. It also became increasingly common for writers to write in their own language—something that had been initiated by the early Italian humanists and gradually spread to the rest of Europe.

The rise in the use of vernacular languages did not change the fact that the universal language of the intellectual world was still

Latin. Printing in Greek never took off in the way Aldus had hoped; there were simply not enough people who knew it to make it a viable proposition for most presses. The inhabitants of the Republic of Letters usually corresponded with each other in Latin, exchanging books and letters, arguing and collaborating via a growing network of postal systems. As the infrastructure of bookselling evolved, it became easier to get hold of texts, helping the exchange of ideas. As the (relatively) stable printed page gradually replaced the fragile manuscript, knowledge became standardized and more accurate. It also became easier to access and consult, as editors and writers added alphabetical indexes, contents pages, diagrams, illustrations and glossaries—the textual paraphernalia we now take for granted.

In 1500, Europe was poised on the brink of the scientific revolution, the seismic discoveries that created the conditions in which science flourishes today. Those discoveries would not have been possible without the centuries of thought, investigation and writing that preceded them, forming continuous threads of knowledge. Captured on the page, scientific ideas travelled around the Mediterranean world, lighting up different places at different times in history. Looking back from our twenty-first-century vantage point, we can see the ebb and flow of this knowledge, the periods of acceleration, and those of stagnation, the ideas that were rejected and lost, only to be rediscovered and revived centuries later. It has not been a straight road, but one that has twisted and turned, run in circles and disappeared down dead ends, before returning once again and moving on.

In recent centuries, technological innovations have transformed scientific knowledge. In the period after 1500, two stand out—both transformed our ability to observe the wonders that surround us. Towards the end of the sixteenth century, a Dutch spectacle maker and his son made primitive microscopes by placing magnifying lenses into a circular tube. One hundred years later, another Dutchman, Anton van Leeuwenhoek, adopted their idea and constructed the first working microscope. He ground and polished 550 individual lenses and placed them inside a tube; this gave a magnification power of ×270, making it possible to see, for the first time, microbes blooming in yeast and blood corpuscles racing through capillaries. From this moment on, unimagined realms of minute detail began to reveal themselves, dramatically altering the intellectual landscape and revolutionizing medicine.

In the early seventeenth century, the astronomer Galileo Galilei took the recently invented telescope, adapted it and turned it towards the night sky. For the first time in the long history of stargazing, it was possible to see beyond the limitations of the human eye, and the universe revealed itself in greater, more wondrous detail than ever before. Since then, increasingly powerful machines have allowed us to see deeper and deeper into the cold recesses of the cosmos and onto the very surface of the moon and planets. Human invention gives us ever-increasing powers of observation, but the more we see, the more appears. Our world is one of seemingly infinite complexity and wonder, revealed to us by science.

ACKNOWLEDGEMENTS

Writing this book has been a long and delightful journey that has taken me back into the distant past and all around the Mediterranean world. Many people have helped me along the way and I am incredibly grateful to them for their guidance and support. First and foremost, my agent, Felicity Bryan, who recognized the germ of an idea and encouraged me to pursue it; and George Morley for her incisive editing, and for guiding me around many pitfalls.

This book is all about scholars in the ancient and medieval worlds, but my ability to trace their journeys and celebrate their achievements is entirely thanks to a galaxy of modern scholars who, through their detailed research and brilliant histories, have made this information available. I have been inspired by many of them, and personally guided by a few. Professor Charles Burnett was extremely generous with his time and unparalleled expertise, his comments on an early draft were invaluable, as were the huge number of books and articles he has written over his distinguished career. He made me feel very welcome at the Warburg Institute and its wonderful library has been my base when I have been in London over the past couple of years. Professor Vivian Nutton was kind enough to share his encyclopaedic knowledge of Galen and answer my questions; others include: David Juste and the Ptolomeaus Arabus et Latinus project, Greg Woolf, Eric Kwakkel, Eugene Rogan, Geri Della Rocca de Candal and Christina Dondi at the fantastic 15c Booktrade Project, Nassima Neggaz and Paolo Sachet for their excellent corrections, Guido Giglioni for his unforgettable Latin teaching, Sabrina Minuzzi for guiding me around the Biblioteca Marciana and fifteenth-century Venice, John Julius Norwich for his kindness in reading my chapter on Sicily. Any mistakes I have made in interpreting their work are entirely my own. The bibliography is not comprehensive. I have included the works that were most important to my thinking and those I believe would be of greatest interest to the general reader.

I have visited many libraries and museums while writing this book and am grateful to all the people who helped me find my way around them: Maria Luz Comendador Perez at the Biblioteca de la Escuela de Traductores de Toledo, Lee Macdonald who showed me the beautiful collection of astrolabes at the Museum of the History of Science in Oxford, Elisabetta Sciarra at the Biblioteca Marciana, Dr. Karen Limper-Herz at the British Library, the helpful staff at the Warburg Library, but most of all everyone at the Bodleian, especially Bruce Barker-Benfield, Colin Harris, Nicola O'Toole, Ernesto Gomez Lozano, Alan Brown, Stephen Hebron and Michael Athanson.

Winning an award from the Jerwood Foundation and the Royal Society for Literature early on in my project gave me a huge boost, psychologically and financially; I am extremely grateful to both organizations and to Molly Rosenberg, Director of the RSL, for her invaluable advice.

I am very grateful to my family and friends for their endless support, patience and ability to continue feigning interest in how the book was going, year after year. Special thanks to Sacha and Adam for telling me to get on with it, Livi and Jenny for making it possible for me to get on with it, JGN for being an excellent research assistant and sometime bag carrier, but also to Dottie, Catherine Nixey, Rob and Charlotte, Cameron, Alexandra, Joanna Kavenna, Thomas Morris, Lucy, Johnnie, Genevieve and Laura, who all inspired and encouraged me along the way. To my parents for sharing their love of history with me. To my girls for putting up with an occasionally absent, and often absent-minded, mother.

Men allermest til Mikkel, for alt.

NOTES

PREFACE

1. Bramante probably advised Raphael on the architectural design, using it to show Pope Julius his vision for the new St. Peter's.
2. However, this figure has also been identified as Archimedes.
3. Owen Gingerich, "Foreword," in G. J. Toomer (trans.), Ptolemy, *Ptolemy's Almagest* (Princeton: Princeton University Press, 1998), p.ix.

ONE: THE GREAT VANISHING

1. Robert Graves (trans.), Suetonius, *The Twelve Caesars* (London: Penguin Books, 1957), Dom. 20.
2. Stephen Greenblatt, *The Swerve: How the Renaissance Began* (London: Bodley Head, 2011), p.106.
3. Choricius, *Laudatio Marciani Secunda* 9, quoted in Averil Cameron, Bryan Ward-Perkins & Michael Whitby (eds), *The Cambridge Ancient History, Volume XIV* (Cambridge: Cambridge University Press, 2001), p.867.
4. Horace Leonard Jones (trans.), Strabo, *Geography* (London: Heinemann, 1932 [Loeb Edition]), 13.1.54.
5. Helmut Koester, *Pergamon: Citadel of the Gods* (Harrisburg, Pennsylvania: Trinity Press International, 1998), p.346.
6. Baynard Dodge (ed.), *The Fihrist of al-Nadim: A Tenth-Century Survey of Muslim Culture* (New York: Columbia University Press, 1970), p.585.

TWO: ALEXANDRIA

1. Horace Leonard Jones (trans.), Strabo, *Geography* (London: Heinemann, 1932 [Loeb Edition]), 17.793–4.
2. Timon of Phlius, quoted in Roy Macleod (ed.), *The Library of Alexandria: Centre of Learning in the Ancient World* (London: I. B. Tauris, 2000), p.62.
3. P. M. Fraser, *Ptolemaic Alexandria* (Oxford: Clarendon Press, 1972), p.133.

4. R. Netz, "Greek Mathematicians: A Group Picture," in C. J. Tuplin & T. E. Rihll (eds.), *Science and Mathematics in Ancient Greek Culture* (Oxford: Oxford University Press, 2002), p.204.

5. Ivor Bulmer-Thomas, "Euclid," *Complete Dictionary of Scientific Biography* (Detroit: Charles Scribner's Sons, 2008), p.415. Hereafter referred to as *DSB*.

6. Ibid.

7. The first two definitions in Book 1, Sir Thomas L. Heath (trans.), Euclid, *The Thirteen Books of The Elements* (New York: Dover Publications, 1956), p.153.

8. Reviel Netz, "The Exact Sciences," in Barbara Graziosi, Vasunia Phiroze & G. R. Boys-Stones (eds.), *The Oxford Handbook of Hellenic Studies* (Oxford: Oxford University Press, 2009), p.584.

9. Gerd Grasshoff, *The History of Ptolemy's Star Catalogue* (London: Springer Verlag, 1990), p.7.

10. G. J. Toomer (trans.), Ptolemy, *Ptolemy's Almagest* (Princeton: Princeton University Press, 1998), p.37.

11. Reliable information on Hipparchus' life is scarce, but he was probably active in Rhodes around 190–120 BC, where he made a series of observations which Ptolemy used in his astronomical models. Although the titles of several works by Hipparchus are known to us, only one has survived.

12. Vivian Nutton, "The Fortunes of Galen," in R. J. Hankinson (ed.), *The Cambridge Companion to Galen* (Cambridge: Cambridge University Press, 2008), p.360.

13. Fridolf Kudlien, "Galen," *DSB*, p.229.

14. "He constructed a systematic and coherent medical synthesis, unparalleled in antiquity in its scope, learning, intellectual aspirations and codification." Christopher Gill, Tim Whitmarsh & John Wilkins, *Galen and the World of Knowledge* (Cambridge: Cambridge University Press, 2009), p.3.

15. Vivian Nutton, "Medicine," in David C. Lindberg & Michael H. Shank (eds.), *The Cambridge History of Science, Volume 2: Medieval Science* (Cambridge: Cambridge University Press, 2013), p.956.

16. Gill, Whitmarsh & Wilkins, *Galen and the World of Knowledge*, p.4. A modern historian put it more succinctly: "The Roman Empire can, with only slight unfairness, be described as overwhelmingly lowbrow in its attitude towards mathematics." A. George Molland, "Mathematics," in David C. Lindberg & Michael H. Shank (eds.), *The Cambridge History of Science, Volume 2: Medieval Science* (Cambridge: Cambridge University Press, 2013), p.513.

17. Vivian Nutton, "The Fortunes of Galen," in R. J. Hankinson (ed.), *The Cambridge Companion to Galen*, p.363.

18. Catherine Nixey, *The Darkening Age: The Christian Destruction of the Classical World* (London: Macmillan, 2017), p.88.

19. Martin Ryle (trans.), Luciano Canfora, *The Vanished Library: A Wonder of the Ancient World* (London: Vintage, 1991), p.192.

THREE: BAGHDAD

1. Quoted in Jacob Lassner, *The Topography of Baghdad in the Early Middle Ages: Text and Studies* (Detroit: Wayne State University Press, 1970), pp.87–91.
2. Michael Cooperson, *Al-Ma'mun* (Oxford: Oneworld, 2006), pp.88–9.
3. Baynard Dodge (ed.), *The Fihrist of al-Nadim: A Tenth-Century Survey of Muslim Culture* (New York: Columbia University Press, 1970), pp.1–2.
4. Nestorian Christians split from the Eastern Church (Orthodox) over doctrinal differences and migrated to Persia and Syria in the fifth and sixth centuries to escape persecution in the Byzantine Empire. They founded monasteries and churches all over the region, many of which remained under Arab rule from the seventh century onwards. Christianity had been widespread in the region since the end of the first century.
5. David C. Lindberg, *The Beginnings of Western Science: The European Scientific Tradition in Philosophical, Religious, and Institutional Context, 600 B.C. to A.D. 1450* (Chicago: Chicago University Press, 1992), p.165.
6. John Alden Williams (trans.), al-Tabari, *The Early Abbasi Empire, Volume I* (Cambridge: Cambridge University Press, 1988), p.143.
7. Ibid., p.144.
8. Paul Lunde & Caroline Stone (trans. & eds.), Mas'udi, *The Meadows of Gold: The Abbasids* (London: Kegan Paul International, 1989), p.33.
9. S. E. al-Djazairi, *The Golden Age and Decline of Islamic Civilization* (Bayt Al-Hikma Press, 2006), p.165.
10. O. Pinto, "The Libraries of the Arabs during the time of the Abbasids," in *Islamic Culture 3*, 1929, p.211.
11. Paul Lunde & Caroline Stone (trans. & eds.), Mas'udi, *The Meadows of Gold*, p.67.
12. This border moved constantly with the vicissitudes of military power, but ran east–west through the centre of what is now Turkey. The wider area was known as Asia Minor or Anatolia, or *Rum* in Arabic.
13. Greek medical manuscripts were part of the treasure won at the battles of Ankara (the capital of modern Turkey) and Amorium, an ancient Greek city in west-central Anatolia that never recovered from the battle and was deserted soon after.
14. There were thirty-six public libraries in Baghdad when the Mongols invaded in 1258.
15. Jafar ibn Barmak was his tutor. It is possible that one of the reasons Harun had him murdered was to break the alliance between al-Ma'mun and the mighty Persian family, reducing his power and making him accept his brother's accession to the caliphate.
16. Baynard Dodge (ed.), *The Fihrist of al-Nadim*, p.584.
17. Robert Kaplan, on *In Our Time: Zero*, BBC Radio 4, 13 May 2004.
18. Jainism is an ancient Indian religion that revolves around several key tenets, including pacifism, celibacy and honesty. Adherents take vows not to steal or

own possessions; they believe in reincarnation and that every living plant and animal has a soul, but they do not believe in any gods. The concept of infinity is central to the Jainist cosmological doctrine, which uses huge numbers to predict planetary and stellar movements far into the future. There are at least 4.2 million Jains living in India today.

19. We can get an idea of the scale of al-Kindi's achievement by the twenty-six separate page references to him in *The Fihrist*, where he is described as, "unique during his period because of his knowledge of the ancient sciences as a whole . . . His books were about a variety of sciences, such as logic, philosophy, geometry, calculation, arithmetic, music, astronomy, and other things." But no one is perfect, apparently, and Ibn al-Nadim completes this impressive list with a final damning comment: "He was miserly." Baynard Dodge (ed.), *The Fihrist of al-Nadim: A Tenth-Century Survey of Muslim Culture* (New York: Columbia University Press, 1970), p.615.

20. Al-Mas'udi, *Murug ad-dahab*, quoted in Dimitri Gutas, *Greek Thought, Arabic Culture: The Graeco-Arabic Translation Movement in Baghdad and Early Abbasid Society (2nd–4th/8th–10th centuries)* (Oxford: Routledge, 1998), p.78.

21. Dimitri Gutas, *Greek Thought*, p.138.

22. Hugh Kennedy, *When Baghdad Ruled the Muslim World: The Rise and Fall of Islam's Greatest Dynasty* (Boston: Da Capo Press, 2005), p.255.

23. Baynard Dodge (ed.), *The Fihrist of al-Nadim*, p.693.

24. Dimitri Gutas, *Greek Thought*, p.138.

25. Introduction to Hunayn's translation of Galen's treatise *On Sects*, quoted in Franz Rosenthal, *The Classical Heritage in Islam* (London: Routledge, 1994), p.20.

26. G. C. Anawati, "Hunayn ibn Ishaq," *DSB*, p.230.

27. Jim al-Khalili, *The House of Wisdom: How Arabic Science Saved Ancient Knowledge and Gave Us the Renaissance* (London: Penguin, 2010), p.75.

28. Baynard Dodge (ed.), *The Fihrist of al-Nadim*, pp.701–2.

29. Information on astrolabes had come to the Arabs in the writings of Theon of Alexandria, whose edition of Euclid's *Elements* they used. The ever-industrious al-Khwarizmi also wrote a manual on how to make and use them, and they became increasingly sophisticated, later being introduced to Western Europe through monasteries in Spain.

30. Colin Thubron, *The Shadow of the Silk Road* (London: Vintage, 2007), p.316.

FOUR: CÓRDOBA

1. Pascual de Gayangos (trans.), Ahmed ibn Mohammed al-Makkari, *The History of the Mohammedan Dynasties in Spain, Volume I* (London: Routledge-Curzon, 2002), pp.17–18.

2. Ibid., p.210.

3. Ibid.

4. Pascual de Gayangos (trans.), Ahmed ibn Mohammed al-Makkari, *The His-*

tory of the Mohammedan Dynasties in Spain, Volume II (London: Routledge-Curzon, 2002), p.81.

5. Pascual de Gayangos (trans.), Ahmed ibn Mohammed al-Makkari, *The History of the Mohammedan Dynasties in Spain, Volume I*, p.121.

6. Paul Alvarus, *The Unmistakable Sign*, quoted in Maria Menocal, *Ornament of the World: How Muslims, Jews and Christians Created a Culture of Tolerance in Medieval Spain* (London: Little, Brown, 2002), p.66. Books available in Latin were mainly on religious subjects and were relatively few in number, meanwhile thousands of books existed in Arabic on a huge range of subjects.

7. Pascual de Gayangos (trans.), Ahmed ibn Mohammed al-Makkari, *The History of the Mohammedan Dynasties in Spain, Volume I*, p.140.

8. Jim al-Khalili, *The House of Wisdom: How Arabic Science Saved Ancient Knowledge and Gave Us the Renaissance* (London: Penguin, 2010), p.196.

9. Sema'an I. Salem & Alok Kumar (trans. & eds.), Sa'id al-Andalusi, *Science in the Medieval World: "Book of the Categories of Nations"* (Austin: University of Texas Press, 1996), p.64.

10. Leon Poliakov, *The History of Anti-Semitism, Volume 2: From Mohammed to the Marranos* (Philadelphia: University of Pennsylvania Press, 2003), p.92.

11. Hroswitha, a German nun who based her description of Córdoba on the testimony of Bishop Recemund. Quoted in Kenneth B. Wolf, "Convivencia and the 'Ornament of the World,'" Southeast Medieval Association, Wofford College, Spartanburg, South Carolina, October 2007, p.5.

12. Hasdai ibn Shaprut, *Letter to the King of the Khazars*, c.960, quoted in Maria Menocal, *Ornament of the World: How Muslims, Jews and Christians Created a Culture of Tolerance in Medieval Spain* (London: Little, Brown, 2002), p.84. The identity of the Gebalim people is uncertain, but they were probably Slavs.

13. Sema'an I. Salem & Alok Kumar (trans. & eds.), Sa'id al-Andalusi, *Science in the Medieval World*, p.72.

14. M. S. Spink & G. L. Lewis (trans. & commentary), Albucasis, *On Surgery and Instruments: A Definitive Edition of the Arabic Text with English Translation and Commentary* (London: Wellcome Institute of the History of Medicine, 1973), p.2.

15. Sami Hamarneh, "al-Zahrawi," *Complete Dictionary of Scientific Biography* (Detroit: Charles Scribner's Sons, 2008), p.585.

16. Sema'an I. Salem & Alok Kumar (trans. & eds.), Sa'id al-Andalusi, *Science in the Medieval World*, p.61.

17. Ibid.

18. Pascual de Gayangos (trans.), Ahmed ibn Mohammed al-Makkari, *The History of the Mohammedan Dynasties in Spain, Volume I*, p.42.

19. Quoted in ibid., pp.139–40.

20. Sema'an I. Salem & Alok Kumar (trans. & eds.), Sa'id al-Andalusi, *Science in the Medieval World*, p.61.

21. Ibid., p.62.

22. Stephan Roman, *The Development of Islamic Library Collections in Western Europe and North America* (London: Mansell, 1990), p.192.

23. Ibid.

FIVE: TOLEDO

1. The preface to Gerard's translation of Galen's *Tegni*, written by his students, translated and quoted in Charles Burnett, "The Coherence of the Arabic–Latin Translation Program in Toledo in the Twelfth Century," in *Science in Context* 14 (1/2) (Cambridge: Cambridge University Press, 2001), pp.249–88.

2. Ibid., p.255.

3. Footnote to Letter 15, Harriet Pratt Lattin (trans. & intro.), *The Letters of Gerbert: With His Papal Privileges as Sylvester II* (New York: Columbia University Press, 1961), p.54.

4. Ibid.

5. Ibid., Letter 138, p.168.

6. The preface to Gerard's translation of Galen's *Tegni*, pp.249–88.

7. A copy of this text, in the version revised by Maslama al-Majriti, with the coordinates adapted to Córdoba, was brought to Zaragoza sometime in the mid eleventh century, where the tables were recomputed to the local latitude.

8. Salma Khadra Jayyusi, *The Legacy of Muslim Spain, Volume 2* (Leiden: Brill, 1992), p.1042.

9. The preface to Gerard's translation of Galen's *Tegni*, pp.249–88 and pp.255–6.

10. Charles Homer Haskins, *The Renaissance of the Twelfth Century* (Cambridge, Massachusetts: Harvard University Press, 1927), p.279.

11. Sema'an I. Salem & Alok Kumar (trans. & eds.), Sa'id al-Andalusi, *Science in the Medieval World: "Book of the Categories of Nations"* (Austin: University of Texas Press, 1996), p.76.

12. Charles Burnett, "The Institutional Context of Arabic–Latin Translations of the Middle Ages: A Reassessment of the School of Toledo," in Olga Weijers (ed.), *Vocabulary of Teaching and Research Between Middle Ages and Renaissance: Proceedings of the Colloquium, London, Warburg Institute, 11–12 March 1994* (Turnhout: Brepols, 1995), p.226.

13. Four of these were versions translated from Greek by Hunayn ibn Ishaq in the ninth century.

14. Vivian Nutton, "The Fortunes of Galen," in R. J. Hankinson (ed.), *The Cambridge Companion to Galen* (Cambridge: Cambridge University Press, 2008), p.364.

15. Angus Mackay, *Spain in the Middle Ages: From Frontier to Empire, 1000–1500* (London: Macmillan, 1977), p.88.

16. Charles Homer Haskins, *The Renaissance of the Twelfth Century*, p.287.

17. Taken from an alternative translation of the preface to Gerard's translation of Galen's *Tegni*, in Edward Grant, *A Source Book in Medieval Science* (Cambridge, Massachusetts: Harvard University Press, 1974), p.255.

18. The preface to Gerard's translation of Galen's *Tegni*, pp.249–88.

19. Peter Dronke, *The History of Twelfth-Century Western Philosophy* (Cambridge: Cambridge University Press, 1988), p.159.

20. Charles Burnett, "The Institutional Context of Arabic–Latin Translations of the Middle Ages," p.225.

21. John of Seville and Limia, who also worked in Toledo, has a background that is somewhat murky. He is referred to by so many different names in the various sources—Hispanus, Hispalensis, Toletanus, Limensis, Avendauth, ibn Dawud—that scholars have wondered whether he may in fact have been more than one person. The current view is that he was probably a Sephardic Jew who fled anti-Semitic persecution in Córdoba under the Almohad dynasty, settled in Toledo and worked on translations there in the mid twelfth century.

22. Charles Burnett, "The Coherence of the Arabic–Latin Translation Program in Toledo in the Twelfth Century," pp.249–88.

23. This is mysterious because Daniel of Morley, who came to Toledo from England in search of knowledge, reports hearing him lecture on the important astrological treatise, Abu Mashar's *Great Introduction to the Science of Astrology*. Providing Daniel was telling the truth, Gerard must have been knowledgeable about the subject, even if he did not translate it.

24. Richard Southern, *The Making of the Middle Ages* (London: Hutchinson, 1959), p.39.

25. Peter Dronke, *The History of Twelfth-Century Western Philosophy*, p.113.

26. Charles Burnett, Hermann of Carinthia, *De Essentiis* (Leiden: Brill, 1982), p.6.

27. David Juste, "MS Madrid, Biblioteca Nacional, 10113 (olim Toledo 98–15)" (update: 01.03.2017), *Ptolemaeus Arabus et Latinus. Manuscripts*, http://ptolemaeus.badw.de/ms/70.

28. Charles Burnett, *The Panizzi Lectures 1996: The Introduction of Arabic Learning into England* (London: The British Library, 1997), p.62.

29. Ibid.

30. Ibid.

31. Charles Burnett, "The Twelfth-Century Renaissance," in David C. Lindberg & Michael H. Shank (eds.), *The Cambridge History of Science, Volume 2: Medieval Science* (Cambridge: Cambridge University Press, 2013), p.380.

SIX: SALERNO

1. Edward Grant, *Physical Science in the Middle Ages* (New York: John Wiley & Sons, 1971), p.4.

2. Cassiodorus, *Institutiones, Book II*, in Leslie Webber Jones (trans. & ed.), Cassiodorus, Senator, ca. 487–ca. 580, *An Introduction to Divine and Human Readings* (New York: W. W. Norton, 1969), p.136.

3. Michael Frampton, *Embodiments of Will: Anatomical and Physiological Theories of Voluntary Animal Motion from Greek Antiquity to the Latin Middle Ages, 400 B.C.–1300 A.D.* (Saarbrücken: VDM Verlag Dr. Müller, 2008), p.277.

4. Ibid., p.304.

5. Ibid.

6. Marcus Nathan Adler (trans.), *The Itinerary of Benjamin of Tudela* (New York: Philipp Feldheim, 1907), p.6.

7. Al-Idrisi, *The Book of Roger*, quoted in Graham Loud, *Roger II and the Creation of the Kingdom of Sicily* (Manchester: Manchester University Press, 2012), p.363.

8. This "head-to-toe" structure was probably copied from Paul of Aegina, the seventh-century encyclopaedist.

9. Lynn Thorndike, *History of Magic and Experimental Science, Volume I* (New York: Macmillan, 1923), p.751.

10. However, as medical teaching developed, Galen's *Art of Medicine* was also included in the *Articella*, so that the two texts could be read in tandem.

11. E. R. A. Sewter (trans.), Peter Frankopan (rev.), Anna Komnene, *The Alexiad* (London: Penguin Books, 2009), p.31.

12. Doctor Pietro Capparoni, *"Magistri Salernitani Nondum Cogniti": A Contribution to the History of the Medical School of Salerno* (London: John Bale, 1923), p.51.

13. Plinio Prioreschi, *A History of Medicine, Volume 5: Medieval Medicine* (Omaha, Nebraska: Horatius Press, 2005), p.232.

14. Faith Wallis, *Medieval Medicine: A Reader* (Toronto: University of Toronto Press, 2010), pp.176–7.

SEVEN: PALERMO

1. Charles Homer Haskins, *Studies in the History of Mediaeval Science* (Cambridge, Massachusetts: Harvard University Press, 1924), p.159, p.191.

2. Prescott N. Dunbar & G. A. Loud (trans.), Amato di Montecassino, *The History of the Normans* (Rochester, New York: Boydell Press, 2004), p.46.

3. Cicero, *In Verrem*, II.2.5., quoted in Dirk Booms & Peter Higgs, *Sicily: Culture and Conquest* (London: The British Museum Press, 2016), p.134.

4. Hugo Falcandus, quoted in Hubert Houben, *Roger II of Sicily: A Ruler between East and West* (Cambridge: Cambridge University Press, 2002), p.75.

5. St. Clement of Casauria, *Chronicon Casauriense*, 889, quoted in ibid., p.75.

6. Hugo Falcandus, quoted in ibid., p.75.

7. This set a precedent—Roger's son, Roger II, also chose porphyry for his tomb, as did successive popes.

8. Many members of the Muslim and Jewish elites had left Sicily during the Norman invasion, so trade with North Africa and the Arab world declined, but did not disappear altogether. The focus shifted to Christian Europe as it began to dominate the Mediterranean. This is nicely illustrated by Ibn Jubayr's voyage from Acre to Sicily in 1184. He sailed on a Genoese ship, along with fifty Muslim pilgrims and 2,000 Christian ones. A century earlier and the proportions would have been reversed, and the ship would most likely have belonged to a Muslim merchant. See: R. J. C. Broadhurst (trans.), *The Travels of Ibn Jubayr* (London: Jonathan Cape, 1952) and Sarah Davis-Secord,

Where Three Worlds Met: Sicily in the Early Medieval Mediterranean (Ithaca, New York: Cornell University Press, 2017), pp.238–9.

9. Al-Idrisi, *The Book of Roger*, quoted in Graham A. Loud, *Roger II and the Creation of the Kingdom of Sicily* (Manchester: Manchester University Press, 2012), p.348.

10. Roger wore the mantle for audiences and to welcome guests, but it was used as a coronation robe by his Hohenstaufen descendants, the Holy Roman Emperors.

11. Hubert Houben, *Roger II of Sicily*, p.121.

12. As described by an anonymous writer in around 1190, quoted in ibid., p.128.

13. Alexander of Telese, *History of King Roger*, quoted in Graham A. Loud, *Roger II and the Creation of the Kingdom of Sicily*, p.79.

14. Jerry Brotton, *A History of the World in Twelve Maps* (London: Allen Lane, 2012), p.73.

15. Al-Idrisi, *The Book of Roger*, quoted in Graham A. Loud, *Roger II and the Creation of the Kingdom of Sicily*, p.357.

16. Hubert Houben, *Roger II of Sicily*, p.98.

17. Part of the so-called *Middle Collection/Little Astronomy* that was studied between *The Elements* and *The Almagest*.

18. *Quaestiones Naturales*, quoted in Charles Burnett, *Adelard of Bath: An English Scientist and Arabist of the Early Twelfth Century* (London: Warburg Institute, 1987), p.10.

19. *Quaestiones Naturales*, quoted in Louise Cochrane, *Adelard of Bath: The First English Scientist* (London: British Museum Press, 1994), p.29.

20. Charles Burnett, *Adelard of Bath*, p.12.

21. Louise Cochrane, *Adelard of Bath*, p.33.

22. Jaqueline Hamesse & Marta Fattori, *Rencontres des Cultures dans la Philosophie Médiévale* (Louvain-la-Neuve: Cassino, 1990), p.94.

23. Charles Burnett, *Arabic into Latin in the Middle Ages: The Translators and Their Intellectual and Social Context* (Farnham: Ashgate, 2009), p.3.

24. R. J. C. Broadhurst (trans.), *Travels of Ibn Jubayr* (London: Jonathan Cape, 1952), pp.339–42.

25. Norbert Ohler, *The Medieval Traveller* (Martlesham, Suffolk: Boydell Press, 1989), p.224.

26. Ibid., p.225.

EIGHT: VENICE

1. Joanne M. Ferraro, *Venice: History of the Floating City* (Cambridge: Cambridge University Press, 2012), p.19.

2. "Poggius Bracciolini to Nicolaus de Niccolis, Letter III," in Phyllis Gordon & Walter Goodhart (trans.), *Two Renaissance Book Hunters: The Letters of Poggius Bracciolini to Nicolaus de Niccolis* (New York: Columbia University Press, 1974), p.26.

3. Stephen Greenblatt, *The Swerve: How the Renaissance Began* (London: Bodley Head, 2011), pp.185–200.

4. *Romeo and Juliet*, 1:4.

5. Konstantinos Sp. Staikos, *The History of the Library in Western Civilization, Volume V* (New Castle, Delaware: Oak Knoll Press, 2012), p.83.

6. C. Doris Hellman & Noel M. Swerdlow, "Peurbach (or Peuerbach)," in *Complete Dictionary of Scientific Biography* (Detroit: Charles Scribner's Sons, 2008), p.477.

7. This manuscript was used as the exemplar for the first printed edition of *The Almagest* in Greek, which was published in Basle in the mid sixteenth century.

8. Paul Lawrence Rose, *The Italian Renaissance of Mathematics: Studies on Humanists and Mathematicians from Petrarch to Galileo* (Geneva: Librairie Droz, 1975), p.48.

9. In 1450, Bessarion had endowed four new professorships in mathematics at the University of Bologna on the pope's behalf, and the two men also commissioned translations of manuscripts by classical mathematicians whose work was little known and who had been overshadowed by Euclid: Diophantus, Apollonius, Proclus, Hero and, above all, Archimedes. Nicholas lent one of these translations of Archimedes to Bessarion. It was never returned, and is still in the Marciana Library in Venice.

10. Bessarion's letter to Doge Cristoforo Moro, quoted in Deno John Geanakoplos, *Greek Scholars in Venice: Studies in the Dissemination of Greek Learning from Byzantium to Western Europe* (Cambridge, Massachusetts: Harvard University Press, 1962), p.35.

11. Lottie Labowsky, *Bessarion's Library and the Biblioteca Marciana* (Rome: Edizioni di Storia e Letteratura, 1979), p.27.

12. Ibid., p.32.

13. Peter Ackroyd, *Venice: Pure City* (London: Vintage, 2010), p.130.

14. In the fifteenth century, there were presses in eighty places in Italy, sixty-four in Germany and forty-five in France. Leonardas Vytautas Gerulaitis, *Printing and Publishing in Fifteenth Century Venice* (Chicago: American Library Association, 1976), p.63.

15. Peter Ackroyd, *Venice*, p.268.

16. Martin Lowry, *The World of Aldus Manutius: Business and Scholarship in Renaissance Venice* (Ithaca, New York: Cornell University Press, 1979), p.191.

17. Ibid.

18. Ibid., p.165.

19. David S. Zeidberg (ed.), *Aldus Manutius and Renaissance Culture: Essays in Memory of Franklin D. Murphy* (Florence: Leo S. Olschki, 1994), p.32.

20. Vivian Nutton, "The Fortunes of Galen," in R. J. Hankinson (ed.), *The Cambridge Companion to Galen* (Cambridge: Cambridge University Press, 2008), pp.367–8.

21. Ibid., p.370.

22. William Eamon, "Science and Medicine in Early Modern Venice," in Eric

Dursteler (ed.), *A Companion to Venetian History 1400–1797* (Leiden: Brill, 2013), p.701.

NINE: 1500 AND BEYOND

1. Neil Rhodes & Jonathan Sawday, *The Renaissance Computer: Knowledge Technology in the First Age of Print* (London: Routledge, 2000), p.1.
2. George Sarton, *Six Wings: Men of Science in the Renaissance* (London: Bodley Head, 1958), p.6.
3. Anthony Grafton, "Libraries and Lecture Halls," in Katherine Park & Lorraine Daston (eds.), *The Cambridge History of Science, Volume 3: Early Modern Science* (Cambridge: Cambridge University Press, 2016), p.240.
4. Elizabeth L. Eisenstein, *The Printing Press as an Agent of Change: Communications and Cultural Transformations in Early Modern Europe* (Cambridge: Cambridge University Press, 1979), pp.567–8.
5. Owen Gingerich, "Copernicus' *De revolutionibus*: An Example of Scientific Renaissance Printing," in Gerald P. Tyson & Sylvia S. Wagonheim (eds.), *Print and Culture in the Renaissance: Essays on the Advent of Printing in Europe* (Newark: University of Delaware Press, 1986), p.55.
6. Thomas Khun, *The Copernican Revolution* (Cambridge, Massachusetts: Harvard University Press, 1957), p.191.

SELECT BIBLIOGRAPHY

PRIMARY SOURCES

Marcus Nathan Adler (trans.), *The Itinerary of Benjamin of Tudela* (New York: Philipp Feldheim, 1907)

R. J. C. Broadhurst (trans.), *Travels of Ibn Jubayr* (London: Jonathan Cape, 1952)

Charles Burnett (trans. & commentary), Hermann of Carinthia, *De Essentiis* (Leiden: Brill, 1982)

H. L. L. Busard, *The first Latin translation of Euclid's "Elements" commonly ascribed to Adelard of Bath*: Books I–VIII and Books X.36–XV.2 (Toronto: Pontifical Institute of Mediaeval Studies, 1983 [Studies and texts])

———, *Campanus of Novara and Euclid's "Elements"* (Stuttgart: Franz Steiner, 2005 Boethius [Series])

———, *The Latin translation of the Arabic version of Euclid's "Elements" commonly ascribed to Gerard of Cremona* (Leiden: Brill, 1984 [Asfār])

———, *The translation of the "Elements" of Euclid from the Arabic into Latin by Hermann of Carinthia (?)*, Books VII–XII (Amsterdam: Mathematisch Centrum, 1977 [Mathematical Centre tracts])

Baynard Dodge (ed.), *The Fihrist of al-Nadim: A Tenth-Century Survey of Muslim Culture* (New York: Columbia University Press, 1970)

Prescott N. Dunbar & G. A. Loud (trans. & eds.), Amato di Montecassino, *The History of the Normans* (New York: Boydell Press, 2004)

Pascual de Gayangos (trans.), Ahmed ibn Mohammed al-Makkari, *The History of the Mohammedan Dynasties in Spain, Volume I* (London: RoutledgeCurzon, 2002)

Phyllis Gordon & Walter Goodhart (trans.), *Two Renaissance Book Hunters: The Letters of Poggius Bracciolini to Nicolaus de Niccolis* (New York: Columbia University Press, 1974)

Edward Grant, *A Source Book in Medieval Science* (Cambridge, Massachusetts: Harvard University Press, 1974)

Mark Grant (ed.), *Galen on Food and Diet* (London: Routledge, 2000)

Robert Graves (trans.), Suetonius, *The Twelve Caesars* (London: Penguin Books, 1957)

Sir Thomas L. Heath (trans.), *The Thirteen Books of Euclid's "Elements"* (2nd ed.: New York: Dover Publications, 1956)

Ian Johnston, *Galen on Diseases and Symptoms* (Cambridge: Cambridge University Press, 2000)

Horace Leonard Jones (trans.), *The Geography of Strabo* (London: Heinemann, 1932 [Loeb Edition])

Leslie Webber Jones (trans. & ed.), Cassiodorus, Senator, ca. 487–ca. 580, *An Introduction to Divine and Human Readings* (New York: W. W. Norton, 1969)

Graham Loud (trans.), *Roger II and the Creation of the Kingdom of Sicily* (Manchester: Manchester University Press, 2012)

Paul Lunde & Caroline Stone (trans. & eds.), Mas'udi, *The Meadows of Gold: The Abbasids* (London: Kegan Paul International, 1989)

O. Neugebauer (trans.), *The astronomical tables of al-Khwārizmi: translation with commentaries of the Latin version edited by H. Suter* (København: I kommission hos Munksgaard, 1962)

Vivian Nutton (ed., trans. & comm.), *Galen: On My Own Opinions* (Berlin: Akademie Verlag, 1999)

———, *Galen: On prognosis* (Berlin: Akademie Verlag, 1979)

Harriet Pratt Lattin (trans. & intro.), *The Letters of Gerbert: With His Papal Privileges as Sylvester II* (New York: Columbia University Press, 1961)

Sema'an I. Salem & Alok Kumar (trans. & eds.), Sa'id al-Andalusi, *Science in the Medieval World: "Book of the Categories of Nations"* (Austin: University of Texas Press, 1996)

E. R. A. Sewter (trans.), Peter Frankopan (rev.), Anna Komnene, *The Alexiad* (London: Penguin Books, 2009)

M. S. Spink & G. L. Lewis (trans. & comm.), *Albucasis on Surgery and Instruments: A Definitive Edition of the Arabic Text with English Translation and Commentary* (London: Wellcome Institute of the History of Medicine, 1973)

G. J. Toomer (trans.), *Ptolemy's Almagest* (Princeton: Princeton University Press, 1998)

John Alden Williams (trans.), al-Tabari, *The Early Abbasi Empire, Volume I* (Cambridge: Cambridge University Press, 1988)

SECONDARY SOURCES

David Abulafia (ed.), *Italy in the Central Middle Ages 1000–1300* (Oxford: Oxford University Press, 2004)

Peter Ackroyd, *Venice: Pure City* (London: Vintage, 2010)

'Aḥmad Azīz, *A History of Islamic Sicily* (Edinburgh: Edinburgh University Press, 1975)

Herbert Bloch, *Monte Cassino in the Middle Ages, Volume I* (Rome: Edizioni di Storia e Letteratura, 1986)

Dirk Booms & Peter Higgs, *Sicily: Culture and Conquest* (London: The British Museum Press, 2016)

S. Brentjes & J. Ren (eds.), *Globalization of Knowledge in the Post-Antique Mediterranean, 700–1500* (London: Routledge, 2016)

Jerry Brotton, *A History of the World in Twelve Maps* (London: Allen Lane, 2012)

P. Brown, *Late Antiquity* (Cambridge, Massachusetts: Belknap Press of Harvard University Press, 1998)

Charles Burnett, "The Coherence of the Arabic–Latin Translation Program in Toledo in the Twelfth Century," *Science in Context* 14 (1/2) (Cambridge: Cambridge University Press, 2001)

———, *The Introduction of Arabic Learning into England, The Panizzi Lectures 1996* (London: The British Library, 1997)

———, *Adelard of Bath: An English Scientist and Arabist of the Early Twelfth Century* (London: Warburg Institute, 1987)

Charles Burnett & D. Jacquart, *Constantine the African and 'Alī ibn al-'Abbās al-Maǧūsī: The Pantegni and Related Texts* (Leiden: Brill, 1994)

Averil Cameron, Bryan Ward-Perkins & Michael Whitby (eds.), *The Cambridge Ancient History, Volume XIV* (Cambridge: Cambridge University Press, 2000)

Luciano Canfora, *The Vanished Library: A Wonder of the Ancient World* (London: Vintage, 1991)

Pietro Capparoni, *"Magistri Salernitani Nondum Cogniti": A Contribution to the History of the Medical School of Salerno* (London: John Bale, 1923)

Louise Cochrane, *Adelard of Bath: The First English Scientist* (London: British Museum Press, 1994)

Roger Collins, *Early Medieval Spain, Unity in Diversity 400–1000* (London: Macmillan, 1983)

Roger Collins & Anthony Goodman (eds.), *Medieval Spain: Culture, Conflict, and Coexistence: Studies in Honour of Angus MacKay* (Basingstoke: Palgrave Macmillan, 2002)

O. R. Constable, *Housing the Stranger in the Mediterranean World: Lodging, Trade, and Travel in Late Antiquity and the Middle Ages* (Cambridge: Cambridge University Press, 2003)

———, *Medieval Iberia: Readings from Christian, Muslim, and Jewish Sources* (Philadelphia: University of Pennsylvania Press, 1997)

M. Cook (ed.), *The New Cambridge History of Islam* (Cambridge: Cambridge University Press, 2010)

Michael Cooperson, *Al-Ma'mun* (Oxford: Oneworld, 2006)

Serafina Cuomo, *Ancient Mathematics* (London: Routledge, 2001)

Sarah Davis-Secord, *Where Three Worlds Met: Sicily in the Early Medieval Mediterranean* (Ithaca: Cornell University Press, 2017)

S. E. al-Djazairi, *The Golden Age and Decline of Islamic Civilization* (Manchester: Bayt Al-Hikma Press, 2006)

Reinhart Dozy, *Spanish Islam: A History of the Moslems in Spain* (London: Chatto & Windus, 1913)

Peter Dronke, *The History of Twelfth-Century Western Philosophy* (Cambridge: Cambridge University Press, 1988)

Eric Dursteler (ed.), *A Companion to Venetian History 1400–1797* (Leiden: Brill, 2013)

Elizabeth L. Eisenstein, *The Printing Press as an Agent of Change: Communications and Cultural Transformations in Early Modern Europe* (Cambridge: Cambridge University Press, 1979)

Joanne M. Ferraro, *Venice: History of the Floating City* (Cambridge: Cambridge University Press, 2012)

Richard Fletcher, *Moorish Spain* (Berkeley: University of California Press, 2006)

Menso Folkerts, *The Development of Mathematics in Medieval Europe: The Arabs, Euclid, Regiomontanus* (Aldershot: Ashgate Variorum, 2006)

———, *Essays on Early Medieval Mathematics: The Latin Tradition* (Aldershot: Ashgate Variorum, 2003)

Michael Frampton, *Embodiments of Will: Anatomical and Physiological Theories of Voluntary Animal Motion from Greek Antiquity to the Latin Middle Ages, 400 B.C.–A.D. 1300* (Saarbrücken: VDM Verlag Dr. Müller, 2008)

Peter Frankopan, *The Silk Roads: A New History of the World* (New York: Knopf, 2015)

P. M. Fraser, *Ptolemaic Alexandria* (Oxford: Clarendon Press, 1972)

L. García Ballester, *Practical Medicine from Salerno to the Black Death* (Cambridge: Cambridge University Press, 1993)

———, *Galen and Galenism: Theory and Medical Practice from Antiquity to the European Renaissance* (Aldershot: Ashgate, 2002)

A. L. Gascoigne, L. V. Hicks & M. O'Doherty (eds.), *Journeying Along Medieval Routes in Europe and the Middle East* (Belgium: Brepols, 2016)

Deno John Geanakoplos, *Greek Scholars in Venice: Studies in the Dissemination of Greek Learning from Byzantium to Western Europe* (Cambridge, Massachusetts: Harvard University Press, 1962)

E. Michael Gerli (ed.), *Medieval Iberia: An Encyclopedia* (London: Routledge, 2003)

Leonardas Vytautas Gerulaitis, *Printing and Publishing in Fifteenth Century Venice* (Chicago: American Library Association, 1976)

Christopher Gill, Tim Whitmarsh & John Wilkins, *Galen and the World of Knowledge* (Cambridge: Cambridge University Press, 2009)

Charles Coulston Gillispie, Frederic Lawrence Holmes & Noretta Koertge (eds.), *Complete Dictionary of Scientific Biography* (Detroit: Charles Scribner's Sons, 2008)

Anthony Grafton (ed.), *Rome Reborn, The Vatican Library and Renaissance Culture* (London: Yale University Press, 1993)

Edward Grant, *Physical Science in the Middle Ages* (New York: John Wiley & Sons, 1971)

Gerd Grasshoff, *The History of Ptolemy's Star Catalogue* (London: Springer Verlag, 1990)

Barbara Graziosi, Vasunia Phiroze & G. R. Boys-Stones (eds.), *The Oxford Handbook of Hellenic Studies* (Oxford: Oxford University Press, 2009)

Stephen Greenblatt, *The Swerve: How the Renaissance Began* (London: Bodley Head, 2011)

Dimitri Gutas, *Greek Thought, Arabic Culture: The Graeco-Arabic Translation Movement in Baghdad and Early Abbasid Society (2nd–4th/8th–10th centuries)* (Oxford: Routledge, 1998)

Jaqueline Hamesse & Marta Fattori, *Rencontres des Cultures dans la Philosophie Médiévale* (Louvain-la-Neuve: Cassino, 1990)

R. J. Hankinson (ed.), *The Cambridge Companion to Galen* (Cambridge: Cambridge University Press, 2008)

Charles Homer Haskins, *The Renaissance of the Twelfth Century* (Cambridge, Massachusetts: Harvard University Press, 1927)

———, *Studies in the History of Mediaeval Science* (Cambridge, Massachusetts: Harvard University Press, 1924)

Lotte Hellinga, *Texts in Transit: Manuscript to Proof and Print in the Fifteenth Century* (Leiden: Brill, 2014)

Hubert Houben, *Roger II of Sicily: A Ruler Between East and West* (Cambridge: Cambridge University Press, 2002)

G. L. Irby-Massie (ed.), *A Companion to Science, Technology, and Medicine in Ancient Greece and Rome* (Chichester: Wiley Blackwell, 2016)

Salma Khadra Jayyusi, *The Legacy of Muslim Spain, Volumes 1 & 2* (Leiden: Brill, 1992)

S. K. Jayyusi, R. Holod, A. Petruccioli & A. Raymond, *The City in the Islamic World* (Leiden: Brill, 2008)

Hugh Kennedy, *Muslim Spain and Portugal: A Political History of Al-Andalus* (London: Longman, 1996)

———, *When Baghdad Ruled the Muslim World: The Rise and Fall of Islam's Greatest Dynasty* (Boston: Da Capo Press, 2005)

Jim al-Khalili, *The House of Wisdom: How Arabic Science Saved Ancient Knowledge and Gave Us the Renaissance* (London: Penguin, 2010)

Thomas Khun, *The Copernican Revolution* (Cambridge, Massachusetts: Harvard University Press, 1957)

Helmut Koester, *Pergamon: Citadel of the Gods* (Harrisburg, Pennsylvania: Trinity Press International, 1998)

Jason König, Katerina Oikonomopoulou & Greg Woolf (eds.), *Ancient Libraries* (Cambridge: Cambridge University Press, 2013)

Paul Oskar Kristeller, "The School of Salerno: its development and its contribution to the history of learning," *Bulletin of the History of Medicine*, Vol. 17 (1945) Feb., No. 2

Paul Kunitzsch, *The Arabs and the Stars: Texts and Traditions on the Fixed Stars, and their Influence in Medieval Europe* (Northampton: Variorum Reprints, 1989)

Lottie Labowsky, *Bessarion's Library and the Biblioteca Marciana* (Rome: Edizioni di Storia e Letteratura, 1979)

Jacob Lassner, *The Topography of Baghdad in the Early Middle Ages: Text and Studies* (Detroit: Wayne State University Press, 1970)

Brian Lawn, *Salernitan Questions* (Oxford: Oxford University Press, 1963)

A. C. Leighton, *Transport and Communication in Early Medieval Europe, AD 500–1100* (Newton Abbot: David & Charles, 1972)

David C. Lindberg, *The Beginnings of Western Science: The European Scientific Tradition in Philosophical, Religious, and Institutional Context, 600 B.C. to A.D. 1450* (Chicago: Chicago University Press, 1992)

David C. Lindberg & Michael H. Shank (eds.), *The Cambridge History of Science, Volume 2: Medieval Science* (Cambridge: Cambridge University Press, 2013)

Martin Lowry, *The World of Aldus Manutius: Business and Scholarship in Renaissance Venice* (Ithaca, New York: Cornell University Press, 1979)

Angus MacKay, *Spain in the Middle Ages: From Frontier to Empire, 1000–1500* (London: Macmillan, 1977)

Roy Macleod (ed.), *The Library of Alexandria: Centre of Learning in the Ancient World* (London: I. B. Tauris, 2000)

M. R. McVaugh & V. Pasche, *Sciences at the Court of Frederick II* (Belgium: Brepols, 1994)

Justin Marozzi, *Baghdad: City of Peace, City of Blood* (London: Allen Lane, 2014)

John Jeffries Martin, *Venice Reconsidered: The History and Civilisation of an Italian City State, 1297–1797* (Baltimore, Maryland: Johns Hopkins University Press, 2000)

María Rosa Menocal, *Ornament of the World: How Muslims, Jews and Christians Created a Culture of Tolerance in Medieval Spain* (London: Little, Brown, 2002)

Elizabeth Nash, *Sevilla, Córdoba and Granada: A Cultural and Literary History* (Oxford: Signal Books, 2005)

Catherine Nixey, *The Darkening Age: The Christian Destruction of the Classical World* (London: Macmillan, 2017)

John Julius Norwich, *A History of Venice* (London: Penguin, 2012)

Vivian Nutton, *The Unknown Galen* (London: Institute of Classical Studies, 2002)
———, *Ancient Medicine* (London: Taylor & Francis, 2004)

Norbert Ohler, *The Medieval Traveller* (Martlesham, Suffolk: Boydell Press, 1989)

Katherine Park & Lorraine Daston (eds.), *The Cambridge History of Science, Volume 3: Early Modern Science* (Cambridge: Cambridge University Press, 2016)

O. Pedersen & A. Jones, *A Survey of the Almagest* (New York: Springer, 2011)

H. L. Pinner, *The World of Books in Classical Antiquity* (Leiden: A. W. Sijthoff, 1948)

O. Pinto, "The Libraries of the Arabs during the time of the Abbasids," *Islamic Culture* 3, 1929

Leon Poliakov, *The History of Anti-Semitism, Volume 2: From Mohammed to the Marranos* (Philadelphia: University of Pennsylvania Press, 2003)

Peter Pormann & Emilie Savage-Smith, *Medieval Islamic Medicine* (Edinburgh: Edinburgh University Press, 2007)

Plinio Prioreschi, *A History of Medicine, Volume 5: Medieval Medicine* (Omaha, Nebraska: Horatius Press, 2005)

R. Rashed & R. Morelon (eds.), *Encyclopedia of the History of Arabic Science* (London: Routledge, 1995)

G. R. Redgrave & E. Ratdolt, *Erhard Ratdolt and his Work at Venice* (London: Bibliographical Society, 1894)

Neil Rhodes & Jonathan Sawday, *The Renaissance Computer: Knowledge Technology in the First Age of Print* (London: Routledge, 2000)

R. T. Risk, *Erhard Ratdolt, Master Printer* (Francestown, New Hampshire: Typographeum, 1982)

E. Robson & J. A. Stedall, *The Oxford Handbook of the History of Mathematics* (Oxford: Oxford University Press, 2009)

Stephan Roman, *The Development of Islamic Library Collections in Western Europe and North America* (London: Mansell, 1990)

Paul Lawrence Rose, *The Italian Renaissance of Mathematics: Studies on Humanists and Mathematicians from Petrarch to Galileo* (Geneva: Librairie Droz, 1975)

Franz Rosenthal, *The Classical Heritage in Islam* (London: Routledge, 1994)

D. F. Ruggles, *Islamic Gardens and Landscapes* (Philadelphia: University of Pennsylvania Press, 2008)

George Sarton, *Six Wings: Men of Science in the Renaissance* (London: Bodley Head, 1958)

———, *Introduction to the History of Science* (Baltimore: Williams & Wilkins, 1927)

George Saliba, *Islamic Science and the Making of the European Renaissance* (Cambridge, Massachusetts: The MIT Press, 2007)

P. Skinner, *Health and Medicine in Early Medieval Southern Italy* (Leiden: Brill, 1997)

Richard Southern, *The Making of the Middle Ages* (London: Pimlico, 1993)

Konstantinos Sp. Staikos, *The History of the Library in Western Civilization* (six volumes) (New Castle, Delaware: Oak Knoll Press, 2004–13)

Lynn Thorndike, *History of Magic and Experimental Science* (New York: Macmillan, 1923)

Colin Thubron, *The Shadow of the Silk Road* (London: Vintage, 2007)

J. V. Tolan, *Petrus Alfonsi and His Medieval Readers* (Gainesville: University Press of Florida, 1993)

S. Torallas Tovar & J. P. Monferrer Sala, *Cultures in Contact: Transfer of Knowledge in the Mediterranean Context: Selected Papers* (Spain: CNERU, 2013)

H. Touati & L. G. Cochrane, *Islam & Travel in the Middle Ages* (Chicago: University of Chicago Press, 2010)

C. J. Tuplin & T. E. Rihll (eds.), *Science and Mathematics in Ancient Greek Culture* (Oxford: Oxford University Press, 2002)

Gerald P. Tyson & Sylvia S. Wagonheim (eds.), *Print and Culture in the Renaissance: Essays on the Advent of Printing in Europe* (Newark: University of Delaware Press, 1986)

Faith Wallis, *Medieval Medicine: A Reader* (Toronto: University of Toronto Press, 2010)

W. M. Watt, *The Influence of Islam on Medieval Europe* (Edinburgh: Edinburgh University Press, 1994)

Olga Weijers (ed.), *Vocabulary of Teaching and Research Between Middle Ages and Renaissance: Proceedings of the Colloquium, London, Warburg Institute, 11–12 March 1994* (Turnhout: Brepols, 1995)

G. Wiet & S. Feiler, *Baghdad: Metropolis of the Abbasid Caliphate* (Oklahoma: University of Oklahoma Press, 1971)

M. Wilks, *The World of John of Salisbury* (London: Blackwell, 1994)

N. G. Wilson, *Scholars of Byzantium* (London: Duckworth, 1983)

———, *From Byzantium to Italy: Greek Studies in the Italian Renaissance* (London: Duckworth, 1992)

——— (ed. & trans.), *Aldus Manutius: The Greek Classics* (Harvard: Harvard University Press, 2016)

David S. Zeidberg (ed.), *Aldus Manutius and Renaissance Culture: Essays in Memory of Franklin D. Murphy* (Florence: Leo S. Olschki, 1994)

INTERNET RESOURCES

A comprehensive guide to books published before 1500: http://15booktrade.ox.ac.uk/

An international project dedicated to the study of Arabic and Latin versions of Ptolemy's astronomical and astrological texts: http://ptolemaeus.badw.de/

A geospatial network model of the Roman world: http://orbis.stanford.edu/

The Bodleian Library's database of manuscripts and images: https://digital.bodleian.ox.ac.uk/

Digitized image bank of the Wellcome Collection: https://wellcomecollection.org/works

INDEX

Page numbers in **bold** denote an integrated illustration.

Illustration Credits

lat.2057 fols.146v–147r © Biblioteca Apostolica Vaticana, reproduced by permission, with all rights reserved)

17. A fourteenth-century Latin copy of *The Almagest* (Science History Images / Alamy Stock Photos)

18. Lavishly painted pages from Roger of Salerno's twelfth-century *Chirurgia* (Sloane MS 1977 fol.6r / British Library, London, UK / © British Library Board / Bridgeman Images)

19. A page of a manuscript of the *Pantegni*, thought to be the oldest medical book in Western Europe (National Library of the Netherlands)

20. Constantine the African lecturing his students (The Bodleian Library, University of Oxford, C 328 fol. 3)

21. Frederick II pictured with one of his beloved birds of prey (MS Pal.lat 1071 fols. 1v © Biblioteca Apostolica Vaticana, reproduced by permission, with all rights reserved)

22. An illustration showing "Saracen" (Muslim), Greek and Latin scribes working at the royal court of Palermo (Burgerbibliothek Bern, Cod. 120.II, f. 101r Credit: Photograph: Codices Electronici AG, www.e-codices.ch)

23. A mosaic in the Norman Palace in Palermo (Palazzo dei Normanni, Palermo, Sicily, Italy / Ghigo Roli / Bridgeman Images)

24. Mosaic in the Martorana Church, Palermo (Granger Historical Picture Archive / Alamy Stock Photo)

25. Roger II's lavishly embellished vermilion silk mantle (DEA Picture Library / Getty Images)

26. Egyptian National Library, Cairo, Egypt / Bridgeman Images

27. A map of Sicily from al-Idrisi's *Book of Roger* (The Bodleian Library, University of Oxford, MS. Pococke 375, fol. 187b-188a)

28. Miniature of Adelard of Bath teaching two students (Leiden University Libraries, ms. SCA 1, f. 1r.)

29. A Christian and a Muslim playing chess (Alamy Stock Photos)

30. Diagram showing phases of the moon, printed by Ratdolt (Historic Images / Alamy Stock Photos)

31. Portrait of Luca Pacioli demonstrating one of Euclid's theorems (Science History Images / Alamy Stock Photos)

32. Opening page of one of Ratdolt's presentation copies of Euclid's *Elements* (British Library, London, UK / © British Library Board. All Rights Reserved / Bridgeman Images)

33. Woodcut of an armillary sphere from a 1543 edition of Regiomontanus' *Epitome of the Almagest* (Granger Collection / Bridgeman Images)

34. A portrait of Poggio Bracciolini (MS Urb. Lat. 224, fol. 2 recto © Biblioteca Apostolica Vaticana, reproduced by permission, with all rights reserved)

35. The manuscript of *The Almagest* in Greek brought to Sicily by Henricus Aristippus, and then acquired by Cardinal Bessarion (Biblioteca Nazionale Marciana, ms Marc. gr. 313 / SHYLOCK e-Solutions di Alessandro Moro)

36. Pages from the Latin translation of *The Almagest* later owned by Colluccio Salutati. (Vat. lat. 2056, fols. 87 verso–88 recto © Biblioteca Apostolica Vaticana, reproduced with permission and with all rights reserved)

37. Galen in an apothecary's shop (Wellcome Collection)

INTEGRATED ILLUSTRATIONS

1. The reconstructed facade of the Library of Celsus in the ruined city of Ephesus (Sorin Colac / Alamy Stock Photo)
2. Map of ancient Alexandria (THEPALMER)
3. Pages from the manuscript of Euclid's *Elements* (The Bodleian Library, University of Oxford, MS. D'Orville 301 f113v-f114r)
4. A reconstruction of the Altar of Zeus at Pergamon (Ullstein Bild / Contributor)
5. Anatomical votives found at the Temple of Asclepius in Athens (*A Companion to Science, Technology, and Medicine in Ancient Greece and Rome* by Georgia L. Irby © 2016 John Wiley & Sons, Inc. Reproduced with permission of the Licensor through PLSclear)
6. A reconstructed map of early Baghdad (Reprinted from Map III from Appendix D, Figures 1, 2, 5, and 6 from Appendix E from *The Topography of Baghdad in the Early Middle Ages* by Jacob Lassner. Copyright © 1970 Wayne State University Press, with the permission of Wayne State University Press)
7. Modern reconstructions of the gates of Baghdad (Ibid.)
8. Map of the Round City (Ibid.)
9. The development and geographical movement of the Hindu-Arabic numerals (Encyclopedia Britannica / Contributor)
10. The Banu Musas' diagram of their self-trimming lamp (Granger Historical Picture Archive / Alamy Stock Photo)
11. An early-eighteenth-century view of Córdoba (Tarker / Bridgeman Images)
12. Reconstructed Arabic water wheel in Córdoba (Alain Machet (2) / Alamy Stock Photo)
13. The Roman bridge over the River Guadalquivir with Córdoba on the left bank (By courtesy of the author)
14. Modern reconstructions of some of al-Zahrawi's intricate surgical instruments (Ibid.)
15. An astrolabe made in Toledo (Granger Historical Picture Archive / Alamy Stock Photo)
16. A fifteenth-century engraving of Toledo (PRISMA ARCHIVO / Alamy Stock Photo)
17. Diagram from a manuscript of Gerard of Cremona's translation of al-Zarqali's *Canones* (Wellcome Collection)
18. A nineteenth-century view of Salerno (De Agostini / Galleria Garisenda- / Bridgeman Images)
19. Robert II of Normandy being treated in Salerno (akg-images / Bible Land Pictures / Z. Radovan / www.BibleLandPictures)
20. A page from the *Circa instans* (Wellcome Collection)
21. A Portuguese map of Palermo harbour (The Protected Art Archive / Alamy Stock Photo)
22. The Norman Palace in Palermo (Michael Wald / Alamy Stock Photo)
23. Illustration of mosaics in the Martorana Church and Palatine Chapel (Florilegius / Alamy Stock Photo)
24. A map of Venice in the twelfth century (Granger Historical Picture Archive / Alamy Stock Photo)
25. An early printing press (Artokoloro Quint Lox Limited / Alamy Stock Photo)
26. A fifteenth-century map of Venice (Age Fotostock / Alamy Stock Photo)
27. First page of Ratdolt's 1482 printed edition of *The Elements* (Granger Historical Picture Archive / Alamy Stock Photo)

28. The Aldine "dolphin and anchor" (Wellcome Collection)
29. A page of Galen's *De methodus methendi* (Add MS 6898, f.1v, British Library, London, UK © British Library Board/All Rights Reserved / Bridgeman Images)
30. Galen dissecting a pig (Wellcome Collection)
31. "Bone man" and "muscle man" from Vesalius' *De humani corporis fabrica* (AF Fotografie / Alamy Stock Photo)
32. Ibid.
33. The Copernican universe (Science History Images / Alamy Stock Photo)